STORAGE
solutions

STORAGE
solutions

Over 100 creative ideas for utilizing space around the home

Cheryl Owen & Phil Gorton

CLB

5054 Storage Solutions

Published in 1998 by CLB International, an imprint of Quadrillion

Publishing Ltd, Godalming Business Centre, Woolsack Way, Godalming,

Surrey GU7 1XW, England

Distributed in the US by Quadrillion Publishing Inc.,

230 Fifth Avenue, New York, NY 10001

ISBN 1-85833-967-7

Printed and bound in Italy

project editor: **suzanne evins**

art director: **simon balley**

design: **balley design associates**

original design concept: **thomas keenes**

jacket concept: **justina leitão**

photography: **jeremy thomas**

stylist: **joanna hill**

illustration: **steve wheele**

US adaptation: **maggi mccormick**

production: **neil randles, karen staff**

& ruth arthur

Acknowledgments

The author and publishers would like to thank Polycell Products Ltd for

supplying the general-purpose putty and Offray Ribbons for the ribbons

used in the projects.

Principles of storage

No matter how large your home is, there never seems to be quite enough storage space. Do you always seem to be endlessly clearing up or hunting for a misplaced item? Then it's worth taking the time to organize your living space and design your own customized home storage solutions.

A growing interest in style and decor in the home means that storage is no longer associated with a recycled shoe box or a stuffed shopping bag. Ideas for storage have become fashionable items for many stores and mail-order companies that sell clever products, often at a high price. Making your own storage items, however, need not be expensive. The projects in this book can be made from fabric remnants and recycled wood, and what's more, they can be custom-made to fit the size of your own rooms and to match your decor. Many of the projects are very versatile and can, of course, be used in other rooms or to store different items to those suggested.

The projects range from very simple ideas that convert ready-made items into designer-style storage systems, such as the tiered vegetable rack (pages 36–37), to handcrafted home accessories, such as the CD stack and speaker stands (pages 78–79). There are projects that will appeal to the novice craft worker that use cardboard, paper, and lightweight plastics, and sewing ideas for those interested in needlework. Some of the ideas in the bedroom chapter are ideal for the beginner, while other sewing projects will inspire the more experienced stitcher.

The woodwork projects are also of varying degrees of difficulty. The adjustable shelving system (pages 54–55) is simplicity itself, requiring a small back or tenon saw, hand drill, and screwdriver and can be attempted by a novice. The blanket chest (pages 78–79) or toy cupboard (pages 102–103) are more complicated and will appeal to the more advanced enthusiast familiar with miters and halving joints. Try your skills out on simpler designs first, and practice the cuts and joints in the Techniques section on scraps of wood before attempting the more difficult projects.

In addition, each chapter focuses on a different room around the home and offers practical and attractive suggestions for streamlining and organizing it to suit your lifestyle and to make day-to-day living simpler. You will find that the key to cutting down on clutter and maximizing your living area is often just a matter of applying some time and thought. Before embarking on any major reorganization, you will have to consider your needs and how they fit in with your present lifestyle, and perhaps more importantly, how your life may alter in the near future. The best storage ideas are those that can be adapted and changed to match new circumstances. You will also need to be realistic about how you actually function; it is no good planning a minimal, clinical environment to live or work in just because the style appeals, if you are basically messy and are happier with your belongings around you.

No matter how busy you are, it is time-effective in the long run to invest some thought as to how to display, disguise, or store your belongings. Once you have worked out a system, stick to it and stop overloading cupboards and shelves and putting items where they don't belong—it will make life much simpler in the future.

Get organized

When faced with a storage challenge, where do you begin? If you are trying to organize a particular area or room, start by deciding what your needs are in that room. Do things just need to be reorganized methodically so you know where to find them without hunting through cupboards and drawers, or would a single custom-made piece of furniture provide all the answers to a range of complex storage problems?

Have a ruthless clear-out—you will be amazed at what has accumulated! You need to eliminate outdated clutter to make room for useful items that are easy to find at a glance. If you are keeping some items "just in case" but have not needed them in the last two years, they probably won't ever be needed and should be thrown out. This often applies to clothes, books, and old music collections. Many of us keep sentimental items such as toys that you or your children have long outgrown. If these mementoes are always packaged away out of sight, and you only see them when you have a clear-out or move, you may want to consider if it's really worth keeping them. A useful habit to adopt is when purchasing a new item for the home, always discard an old one; this keeps your environment fresh and does not encroach on existing space.

Consider how often you need access to your things. Keep items that are used most frequently at hand or eye level so you can get your hands on them with minimal effort, and store those items which are needed only occasionally within your furthest reach. Heavy objects that are used less frequently, such as a sewing machine or a boxed vintage record collection, need to be kept between knee and floor height.

Seldom-used items—for example, suitcases, decorating materials and Christmas decorations—can be stored in less accessible areas such as in the attic, a cellar, or on top of cabinets. Store small items to go in an attic or cellar in bags or lightweight boxes so they are easy to maneuver (see the corrugated-plastic boxes on pages 122–123). Bags should be sealed, and boxes should be closed with lids so the goods do not become dusty. Label all your items clearly to avoid frustration at a later date! If dampness is likely to be a problem, store articles in sealed plastic bags first. Sometimes, slight cellar damp can be rectified by a dehumidifier, releasing a large area for storage that may have be considered unusable. If you have attic space, consider boarding it with flooring-grade board and attach a foldaway ladder; not only will this be safer than climbing on chairs and freestanding ladders, but the ease of access will undoubtedly encourage you to use the attic space more often. Remember when you are storing items in an attic not to block roofing vents, for a badly ventilated area will quickly become a damp one.

Before making or purchasing new furniture or shelving systems, it is a good idea to map your room to scale on graph paper. Use a metal tape measure to measure the room; a fabric one may stretch and distort. Measure all the walls—do not assume that they are symmetrical. Draw a rough sketch, making

a note of the wall measurements, position of windows and doors, and which way they open. Write down the height of the room, add the baseboard, and note any jutting architectural features such as fireplaces, the existing furniture, radiators, and electrical outlets.

Now draw the room to scale on graph paper. For example, on 1:48 graph paper each small square equals 6 inches. Mark on the existing features. In the same way, draw the elevation of the walls on another sheet of graph paper. Tape a sheet of tracing

above: **To maximize storage in your living space, first make a draft plan of the room, including furniture, doors, windows, and fireplace, before drawing the plan to scale on graph paper.**

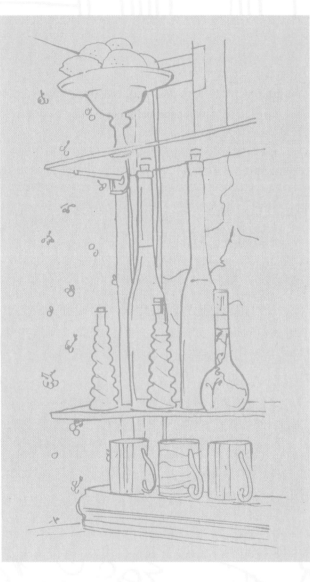

right: **A glass shelf fitted across a bathroom window provides an attractive display area for glass bottles, and also provides some privacy, while at the same time allowing light into the room.**

Lining shelves and cupboards up with existing features such as tops of doors and windows make the new pieces appear integral to the room. You may want to move sockets to make them more accessible for lighting or using electrical equipment. Continue the baseboard and moldings, if applicable, around the structure if you are building in a storage unit.

Take time to consider whether your proposals work practically on the room and elevation plans, then see if they are also pleasing to the eye before starting any work. Think about the lighting, too, since it will be affected by major alterations. Ideally, the lighting in a room should be harmonious and not dazzle you or throw dark shadows into areas of activity. Concealed strip lights can be used to highlight display areas, and interior lights in cabinets can be controlled by opening and closing the doors.

If you look at your home carefully with regard to storage, you will see new possibilities with unused "dead" space. Glass shelves across a bathroom window will let in the light while providing privacy and can hold toiletries or display pretty glass pieces. The space above a door can be fitted with a shelf to hold books or vases, and a narrow shelf can be added to a picture rail. Put shelves in a non-functioning fireplace or attach doors. A built-in window seat in a bay window not only adds extra seating, but can be designed with chests underneath to provide a large storage area. Shelve an alcove or awkward, small corner of a room, maybe with a cupboard below or hang a blind or curtain from the bottom shelf to hide clutter.

Shelves and shelving

paper on top and draw on your planned storage. Remember to allow clearance of doors and allow access to windows and electrical sockets and of course to the storage itself. Draw alternatives on other sheets of tracing paper.

While you are putting together your plan, bear in mind that a single span of shelving or cupboards is more attractive and more useful than a few smaller pieces dotted around the room.

You will find shelving of some sort in every home, since it is the most flexible type of storage. A shelving system is not only practical, since it can be built into "difficult" corners; it can also be as basic and inexpensive—or as sophisticated—as you wish. Adjustable shelves are the most popular since they can adapt to your changing needs, and so are particularly useful in children's rooms. They are supported by dowels or wooden blocks that slot into holes in wooden struts or brackets, or clips that slot into slits in metal rods that are attached to the wall. The height of the shelf is determined by the hole or slit, so it can be changed if necessary. Stationary shelving is often stronger since it is supported on wood strips or brackets. The brackets can be a design feature themselves. Concealed fixtures take the form of D-shaped brackets that are mounted on the wall and slotted into grooves in the ends of the shelf.

A recess on either side of a fireplace is a favorite spot for shelving, since it can become an attractive feature while at the same time making practical use of what is sometimes a difficult part of the room to furnish. For any heavy storage items, consider not only battening the ends of your shelf, but the back also. A support on the back wall will make the shelf extremely strong, and by running an identical wood strip on top of the shelf, you can line the back wall with a decorative wooden backing such as upright tongue-and-groove strips, or make plywood panels with architrave surrounds between the shelves. This creates a miniature environment reminiscent of old-fashioned paneled libraries.

All shelves must be well supported to prevent unpleasant accidents to persons and property—the heavier the items on the

left: The shallow alcoves on each side of a fireplace provide an ideal place for extensive shelving. Stacked books and family photographs enhance the atmosphere of the room.

above: **A simple drawer organizer which can be made from scraps of cardboard or thin plastic to hold small items.**

right: **Plain shelving can be made into an attractive feature by attaching decorative edging strip.**

shelves, the more supports needed. Bear in mind that you will probably store more on them over the years. The thickness of the shelf must be relative to the span, and it must be able to hold the contents without sagging.

Shelving can be constructed from either planed soft woods or cut to size from man-made boards. Soft woods are available as linear lengths from hardware stores or lumberyards. When cutting manmade boards to length for shelving, use ¾-inch boards. Always have the board cut so the grain is along the shelf, never cut across the grain. A popular choice for shelves is 1-inch thick linear wood.

Cheaper soft woods such as deal can be stained, varnished, or painted. Man-made boards such as blockboard and plywood with an external veneer such as birch are easily stained and finished, but they need to be edged with a matching edging strip; ask your wood seller about this when buying the board. For a more sophisticated look, an oak-veneer finish is more expensive and can be polished. Materials such as melamine and other plastic-coated boards are inexpensive and easy to clean, but they have little stylistic quality.

The most economical shelving is machine-cut in lengths from board. Ideally, select a lumberyard that will cut the board for you, pay for the entire board, and keep the spare pieces if they are useful sizes. At the same time, purchase edging strips in the same wood as the main veneer surface (a lot of boards are one-sided) for a professional finish. Edging strips are also available in hardwood—ramin, a hardwood from Indonesia and West Malaysia is very popular. Edging strips can be used as a contrasting color, since they finish as a white wood that is very effective on a redwood board. Be creative with molded finishes, styling your shelves with decorative edging patterns for a distinctive look.

Glass can be used for shelving, but it is not as versatile. Toughened glass shelving is expensive, but looks very attractive if well lit for display purposes. It will, however, only support lightweight items.

Drawers and cupboards

Organizing drawers sensibly can go a long way to creating extra space. There is inevitably a tendency to overfill drawers, making them heavy to maneuver and difficult to sort. Try to keep drawers no more than two-thirds full and spray the runners with furniture polish to help them run smoothly. Follow similar principles for packing a suitcase when you are putting clothes into a drawer—fold items as flat as possible to make them less bulky, and keep things that you wear most often near the top of the drawer.

To house small items, choose a shallow drawer and divide it into compartments. Recycle smaller plastic containers and place them in the drawers, or make your own from wood or thin plastic to fit the drawer exactly and divide it into suitable sections. Alternatively, long strips of cardboard or thin plastic can be glued or clipped together to create simple and effective drawer organizers for small toys or socks.

A stand of wire baskets or transparent plastic crates can make a stylish alternative to a chest of wooden drawers, and the contents are always visible to minimize hunting. Square or rectangular wicker baskets become improvised drawers when they are stacked on shelves, and they can also be lifted out easily for added portability. Wire and wicker baskets allow air to circulate, so they are always ideal for both food and clothing.

Small, unpainted and unvarnished wooden chests are available inexpensively from selected home furniture stores. They are ideal for any room in the house and can be stained, painted or decorated flamboyantly to house small items that could otherwise go astray.

Blinds, curtains, and doors

Many stored items need to be concealed or disguised, either to give a streamlined look to the surrounding room or for protection. One of the best ways to do this on a freestanding shelving unit is to put a window shade to the top to roll down and hide clutter, or try a venetian blind for recessed shelving in an office setting.

Curtains can be hung in front of an open cupboard or table from curtain wire or threaded through dowel on eyelets. Support the dowel at each end on hooks and simply lift off for access. Fabric attached behind glass- or wire-fronted doors softens the look and gives a country feel to a room. Glass-fronted doors can be decorated with an etching kit or painted with glass paints to hide the contents.

Fold-back or sliding doors are suited to tight spaces and are ideal for closets or built-in cupboards. Don't neglect to use the backs of doors for hooks or rods to hang coats, bags, or towels. For a less permanent disguise, consider investing in a screen; their potential versatility in any room is underestimated. Whether they are made from canvas attached to a simple frame, or from three hinged and painted pieces of medium-density fiberboard, they can merge into the background or be a bold statement while hiding clutter.

below: **Portable wicker or wire baskets make a storage alternative to traditional cabinets and drawers. Wire baskets have the additional advantage of keeping items visible.**

right: **Hide unsightly housework equipment out of the way when not in use. This iron and board have been hung from a special wall attachment.**

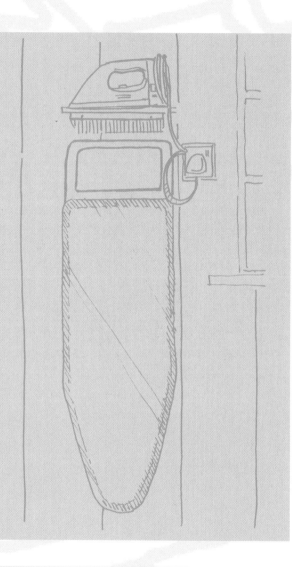

the sink or bathtub to drip-dry clothes or choose one to hook over a radiator that can be removed when not in use. Store a pile of ironing in an attractive wicker basket, so it can be lifted up and tucked away out of sight when not being tackled.

The space under the stairs is often overlooked when planning storage, but it can be shelved by extending the stair treads backward or turned into a spacious cupboard for numerous practical purposes. This could become a wine cellar with built-in wine racks; a functional cupboard for cleaning materials, the vacuum cleaner, and ironing board; or a coatroom with hooks for coats, a shelf for hats and gloves, and a rack for boots, as well as a place to store a folding stroller. To maximize the space available, it is wise to first plan the storage in a large cupboard area by measuring and then making a list of the items that need to be stored, so that everything can be stacked, hung, and shelved in its place. Install a small light inside the cupboard so that you can see into the deepest recesses; small battery-operated lights can easily be attached with a clip or a touch-and-close fastening.

Remember to designate space to house recycling materials, which can look unsightly. At the very least, have a stash of shopping bags to slip empty bottles into and a rack for newspapers and magazines. Use waste bins that are designed with separate sections for metals, paper, glass, and plastic or devise your own system by using stackable plastic crates for different waste materials.

Utility areas

If you don't have a room in the home designated for housework and laundry alone, they must be tackled in other areas as unobtrusively as possible. Choose a washer-dryer rather than two machines, or a small-size washing machine that can have a tumble dryer stacked on top. Opt for special fixtures that can be attached to the inside of doors to hold vacuum cleaner parts, irons, and ironing boards. Position a pull-out clothes rack above

Innovations

Storage does not need to be purely utilitarian—why not have some fun and devise a few original storage systems that grab attention? Suspend shelves on ropes or chains, or stack them using a matching set of sturdy ceramics for support. Use square glass vases or fish bowls to support shelves of ornaments. Continue the theme of your display by placing some of the objects inside the glassware; this will stop them being from handled and protect them from dust.

To make a flamboyant children's shelving system that resembles a biplane, cut a large plastic water bottle in half. Insert a brass paper fastener outward through the cap and tape it in position on the inside. Screw on the cap and cover the bottle with a few layers of papier-mâché, then paint it to look like the front half of the plane, complete with a pilot. Cut a propeller from cardboard and push the prongs of the paper fastener through the center. Glue the "plane" between two shallow shelf "wings".

Salvage old metal containers or food cans, remove any paint and rust, then apply color with enamel spray paints or leave them in their natural state for a fashionable metallic finish. Some packaging can be recycled to make very attractive containers. Wooden wine and liquor boxes with their stamped logos look very stylish wood-stained or varnished. Stand them upright and add a door to the front to make a small cabinet. Printed metal cans, such as those used for cooking oil or Chinese tea, make great containers for utensils in the kitchen. Use a can opener to remove the top of the oil can and file the cut edges to smooth them.

Signal the contents of your storage with some extrovert labeling. Photograph the contents and tape the photograph to the container. Children will love fun storage! Paint images of the contents on drawers and cupboards in a young child's bedroom. Clearing up will become a game as the youngster matches the items to the pictures. Draw a coat hanger on corrugated cardboard, making the hook into the head of a bird. Cut out the shape and build it up with papier-mâché, then paint it to resemble a fantasy bird of paradise. Glue a squeaker (available from toy-making suppliers) to a drawer or door knob, and cover the whole thing with bright fur fabric to give a vocal surprise when the drawer or cupboard is opened!

left: The area under the stairs can often be converted into an ideal storage space. In this room the alcove has been shelved to hold decorative display items.

below: Create a whimsical biplane shelf for a child's room using a recycled plastic bottle, papier-mâché, and two small wooden shelves.

Materials

You may already have some of the materials needed for the projects, but most items are readily available from art, craft, fabric, and hardware stores. It is always worth choosing good-quality materials; some may cost a little more, but your finished project will last longer and look better.

Fabric crafts

Closely-woven fabrics are the most versatile for the fabric projects. Canvas ❶ is inexpensive and is available in various widths. It handles well and has a crisp finish when pressed. Plastic-coated fabrics ❷ are water-resistant, so they are ideal for items that will be used for travel or the bathroom. Transparent plastic, which is available from upholstery departments and craft stores, is a great material for making pockets and containers since the contents will be visible.

Finish raw fabric edges with bias binding ❸, or if you wish to make your own, refer to the technique on page 24. Use metal D-rings ❹ to attach fabric tapes. Make bag handles from lengths of strong cotton webbing and use cotton tape for tie fastenings. Glue fabric with latex glue or craft glue ❺. Craft glue can also be applied to fabric to stiffen it and prevent fraying. Plastic boning ❻ is a flexible plastic strip used for corsetry. Buy it from the notions counter for stiffening the tops of fabric containers. Notions counters sell continuous zippers ❼, which are ideal for the large fabric projects in the bedroom chapter. The zipper is cut to your chosen length with a pull on it. The ends can be taped or handsewn together until the zipper is used, so that the pull does not slip off the unfinished cut ends.

Touch-and-close fastening ❽, such as Velcro, comes in various colors and widths. It consists of two strips of tape, one with a soft, looped surface and the other with tiny plastic hooks. Press the surfaces together to fasten them and simply pull apart to undo. Stitch close to the long edges to attach to fabric or apply to wood with brads. Adhesive-backed versions are also available. Metal snaps ❾ and eyelets ❿ give a professional finish and can be bought in kit form.

Craftwork

Wire ⓫ comes in various gauges and can be manipulated with pliers. Cut wire with wire snippers. Chicken wire ⓬ is cheap to buy and will give a rustic feel to a project.

Recycle thick cardboard from packaging to make containers to be covered with fabric. Large boxes made from corrugated cardboard ⓭ can be cut up and used to stiffen inside fabric items. Corrugated plastic ⓰ is available at art and specialist craft stores. Mat board is made of layers of cardboard and foam board **14** consists of a layer of foam sandwiched between two layers of cardboard. They are both available at art supply stores. Craft stores sell thin, lightweight plastic **15** and colored foam that are great fun for craftwork. Cardboard, foam, and plastic materials can be joined with metal paper fasteners **17** to give a professional and stylish finish.

Always read the manufacturer's instructions for adhesives and test them on scrap paper before use. A plastic spreader or a piece of cardboard is useful for distributing glue evenly. Craft (polyvinyl acetate) glue **5** is a versatile nontoxic solution that dries to a clear finish. Spray adhesive gives an even coat of glue on papers and boards, and is useful if you need to glue a large surface, but it must be used in a well-ventilated area. Keep a tube of all-purpose household glue **18** on hand, as it is used for many of the projects. Paper glue is not particularly strong but will stick paper to board.

Double-sided tape **19** is a clean alternative to glue. The tape is sticky on both sides and has a backing paper that can be peeled off when ready for application. Masking tape **20** is a low-tack tape, which means it can be used to hold paper and templates in place temporarily and removed without marking the surface. General-purpose marking tape **21** is useful for labeling and binding components together while you work.

Woodwork

The wood projects in this book are made from a number of sources: blockboard **22** and plywood **23** which are man-made sheets, made from blocks and strips of real wood respectively, with an external veneer; softwoods **24** which are sold in lengths; hardwoods such as ramin edging strip **25**; and dowel rods **26** which are available in various thicknesses. Sheet materials are sold in standard sizes. They can be cut to size by a lumberyard or hardware store and then bought by the square foot or square yard.

Softwoods are sold in nominal size units—nominal because they are referred to in their sawn sizes. It is important to remember that when the wood is planed and "prepared all around" the waste wood is not deducted from the finished size. Therefore 1 x 2-inch wood actually measures slightly less.

Wood glues are fast-acting and easy to apply because they are sold with a handy directional nozzle. Green-colored packaging **27** denotes regular adhesive, and waterproof glue is found in dark-blue containers **28**. Grab gap filling adhesive is an extra-strong glue often used for woodwork.

Buy screws **29** and seats **30** from a hardware store. Most retailers will price them in multitudes of ten, but always overestimate the number you require in case of badly machined slots or other faults. The projects use countersunk head slotted self-color screws with brass seats where needed. A small amount of grease on the thread eases the passage of the screw ❼❹. Where the heads are countersunk, fill with plastic wood ❸❶, which comes in various finishes, natural or pine being the most common. General-purpose putty ❸❷ is acceptable if the item is to be painted. Apply the putty with a knife or trowel to the holes and any defects.

Woodworking nails are sold in small packs. For the projects, use 1-inch brads ❸❸ and 1-inch molding tacks ❸❹ on the edgings. Molding tacks are thinner than brads and therefore will not split small strips of wood, as the larger brads will. Hold the tacks in place with a piece of cardboard with a slit in it when tapping them in.

Sandpaper ❸❺, or more accurately glass paper, comes in coarse, medium, and fine grits. When using sandpaper, tear the sheet into quarters; never cut it with scissors or a knife as a sharp edge will damage the wood surface. Start sanding with coarser paper and move down to medium and fine grits to effect a finish, eventually using a fine sandpaper wrapped around a block of wood ❸❻. Key the surface between coats of paint or varnish by sanding with fine sandpaper, working in the direction of the wood grain.

Cup hooks come in various styles. Stick-on hooks are strong enough to hold lightweight items, but screw-in hooks ❸❼ are more versatile and longer-lasting. They are available in different sizes and finishes such as brass, self-colored, and white plastic-coated. Use S-hooks ❸❽ to suspend metal chains.

If you wish to paint your project with an oil-based paint, apply a suitable primer first. Water-based latex paints are fast-drying, but it is a good idea to apply a couple of coats of clear varnish to the finished piece for protection. Spray paints are quick and fun to use and give an even finish. They should be used in a well-ventilated room or outdoors, with the surrounding area protected with lots of newspaper. Always wear a face mask and use sprays that are free of CFCs.

Bare wood looks very stylish with a few coats of clear varnish. See the Finished! technique on page 29 for making a wood stain if you want to color the wood. Ready-colored varnishes are available, but do not give such a good-quality finish as a handmade stain. Varnishes come in gloss, satin (semigloss), and flat finishes. Gloss is the most hardwearing finish and flat the least. For a durable flat finish, apply gloss varnish first, then the flat varnish. Water-based acrylic varnish dries quickly to a clear finish. Polyurethane oil-based varnish takes longer to dry and has a yellowing finish, which is effective if you want an antique feel to the piece.

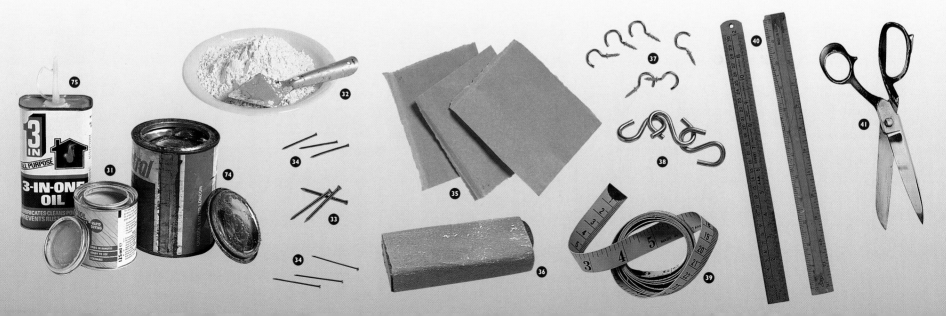

Equipment

Always use the right tool for the job; improvising may take longer and give a poor finish. A basic sewing kit and sewing machine is all that you will need for the fabric projects, and desk and drawing equipment for the craftwork. A basic tool kit for household repairs is a vital part of the home, and you will be able to use items from here for most of the wood projects. An additional small, specialized woodworking kit will be needed and the small handtools required are detailed below. Power tools are not essential for any of the projects, and although their use will speed up the working process, bear in mind that speed and ease of use are no substitute for accuracy. Remember that any board-cutting can be done by the lumberyard.

Fabric crafts

Use a fabric tape measure ❸❾ for measuring fabrics and draw straight lines with a ruler ❹⓿. Lines drawn with an air-soluble pen (available at craft and fabric stores) will fade away after some hours, making them ideal for marking fabric. Always test first on a scrap of fabric.

Use dressmaking scissors ❹❶ to cut fabrics and a pair of embroidery scissors ❹❷ for snipping threads and fabric, keeping them sharpened regularly. Do not use fabric-cutting scissors to cut paper or the blades will become blunt quickly. Pin fabrics

with glass-headed dressmaking pins, which will show up clearly against the fabric ❹❸. Use the correct-size hand and machine needles for the fabric and thread you are using. A staple gun ❹❹ is useful when covering wood with fabric.

Craftwork

Use an hard pencil ❺❷ for drawing, keeping it sharpened to a point for accuracy, or use a propelling pencil. Use a ruler and drawing square ❹❺ when drawing squares and rectangles so the lines are straight and the angles accurate. Use a drawing compass to describe circles.

Sharp scissors ❹❻ are indispensable for craftwork. A craft knife ❹❼ or scalpel gives a neat cut on paper, cardboard and plastic. Change blades regularly since a blunt blade will tear the surface of paper and board. Always use a craft knife on a cutting mat. Plastic cutting mats are available from art and craft stores. They have a self-healing surface so can be reused repeatedly. An awl ❹❽ is useful for making holes in plastics.

Woodwork

Measure along boards and linear lumber with a metal tape measure ❹❾ and use a conventional ruler ❹⓿ in either wood or steel for smaller measurements. Square up your measurements with a carpenter's square 78 or drawing square 45 and prepare for miter cuts using a combination square 50 (some have an integral carpenter's level). Mark boards for cutting with a carpenter's pencil 51, and joints and small cuts with a marking knife 53 or craft knife 47. Cut large boards with a general-purpose rip saw 54. Sawing is made easier with a drop of oil on the blade 75. Cut linear boards to size with a medium-size tenon saw 55 and cut joints or make restricted access cuts with a small back saw 56. Use the tenon or back saw supported on a bench hook 57 for 90 degree cuts. Miter 45 degree cuts in a miter box 58.

Sink brads and molding tacks with a small pin hammer 59. Use the heavier claw hammer 60 to counter-punch the nail heads below the surface using the nail punch 61. Use awls 48 to make starter holes for small screws, using an appropriate screwdriver 62 to drive them. Larger screws need for a hole to be drilled first. Use a twist bit 63 and center punch 64 or a self-

starting auger bit in a brace 65, countersinking if desired by using a countersink bit in a hand drill 66. Larger holes can be drilled with a flat bit 76.

Wood waste in rabbets can be taken out with a mortise or firmer chisel 67, driven by hand or a wooden mallet 68. Hold the work piece steady with a C-clamp 69 or a bar clamp 70. When using clamps, protect the project with small scraps of wood 77. Make sure all blades are sharp by using an oilstone 71. Remove old nails with a pair of pincers 72.

A badger-hair brush is best for applying varnish, but a good-quality paintbrush 73 would also be suitable. Clean paintbrushes immediately after use. Use mineral spirits for oil-based paints and varnishes. Mineral spirits can also be used to wipe away pencil marks on wood.

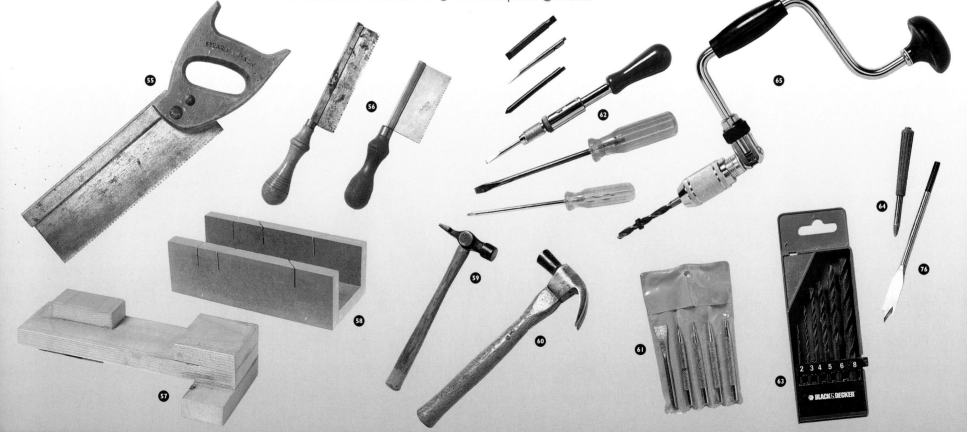

Safety first

Few people have the luxury of a separate workshop in the home. Consequently, your work space when making the projects is often temporary and also used by others, including children. Safety is therefore a prime consideration.

Make sure the floor is free from obstruction. Sockets are generally at baseboard level, so if you are using power tools, do not leave them on the floor with trailing cord where you are liable to trip over them. An extension-cord or box allows you to leave power tools on the worktable, out of the way. Do not allow cords to intertwine, do not change drill bits with the power on, and never pick up power tools by the cord. Remember that wood burns, so keep an extinguisher in the workplace. Dust and shavings should be swept up into a metal trash can with a lid, wearing a face mask while doing this.

Hand tools such as saws and chisels have sharp teeth and edges. Protect saw teeth with a tooth guard, which is usually supplied with good-quality tools; protect chisels with plastic tip guards. If the guards are lost, wrap the blade in cardboard and secure it with tape. When using sharp tools such as chisels, always cut or chisel away from your body or support hand, considering your personal safety at all times. Never leave craft or marking knives in a drawer with the blades exposed; anyone hunting through the drawer contents could receive a nasty cut. Make sure a clearly identified first aid kit is in the workplace.

Never hang power tools up. Put them away safely in a cupboard or secure box. If you use an extension-cord, never leave the tools connected or the reel switched on when not in use. Always work methodically and clear up as you go; do not leave it to the end of the day. Secure the lids on adhesives, paint, and varnish to prevent them from drying out or spilling. Put screws, nails, and dressmaking pins back in their containers and always replace tools in their proper place; this means they can always be found and not cause injury. A neat workplace is also an efficient workplace.

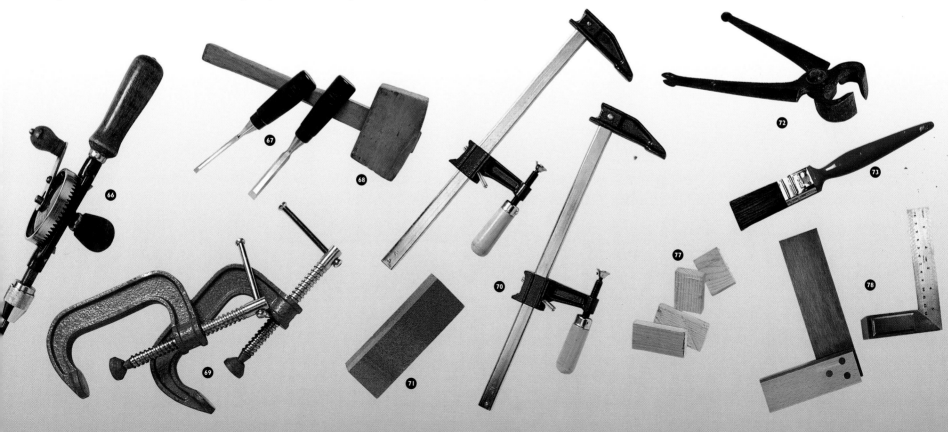

Techniques

THIS SECTION of the book outlines the basic techniques that are used in many of the following projects. Read the instructions carefully before embarking on any of the projects, and refer back to these pages if you come across an unfamiliar technique.

Bias strips

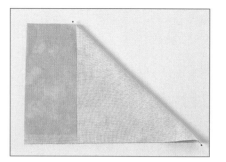

1 The bias is the diagonal 45 degree angle to any straight grain of fabric. Strips of fabric cut on the bias are used to bind raw edges of fabric. To find the bias, lay the fabric out flat and make sure the edges are straight. Fold one edge to meet the edge adjacent to it. Mark with a pin at each end.

2 Open the fabric again and use a ruler to draw a line between the pins. Draw another line parallel to the first for the width of the bias binding, include seam allowances and double the width for the binding being used double, if necessary. Cut out the strip, then stitch a number of strips together to make a longer length. Ready-made bias binding is available in various widths. The width of ready-made bias binding given in the projects does not include the seam allowances.

Finishing seams

In order to finish seams and prevent them from fraying, work a zigzag stitch along the raw edges. Alternatively, trim the raw edges with pinking shears.

Clipping curves and corners

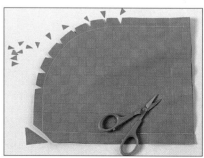

To reduce the bulk in curved seams, clip V-shapes in the seam allowance almost to the stitching. Clip across the seam allowance at the corners so the fabric will lay flat when turned right side out.

Measuring and marking

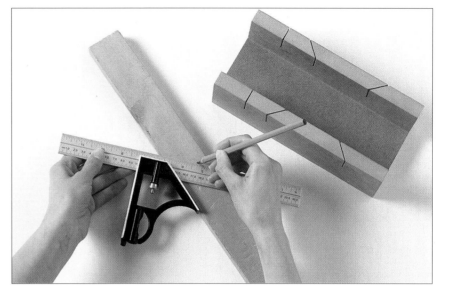

1 Take time to measure accurately. It is essential when transferring a template or diagram to cardboard, wood, or fabric to line it up first on a straight edge. Using a drawing square against a (preferably) finished cut edge will start the drawing with a 90 degree angle. Continue this technique to form a square or rectangle.

2 When working with wood, mark cuts to be made with a crosscut saw using a carpentry pencil. Saw the cut in the middle of the line. Blow sawdust away from the teeth of the saw so it does not get in the way. Mark accurate cuts to be made with a tenon saw using a marking knife or craft knife. When marking a wood joint, use one half as a template to mark the other, do not use a measure. Always cut on the waste side of the line.

3 In order to prepare miter joints for cutting using a miter box, first mark the wood with a combination square at an angle of 45 degrees. The combination square is a useful tool, as it can function as an amalgam of tri-square, straightedge, miter rule, gauge, and carpenter's level.

Using a craft knife and scalpel

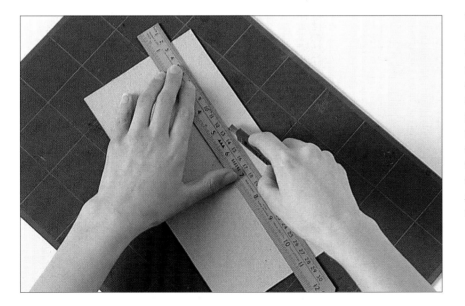

Work on a cutting mat set on a flat, stable surface. Do not press too hard or attempt to cut right through the material you are using at the first approach, but gradually cut deeper and deeper. To score cardboard or plastic, do not cut right through the material, but break the top surface only so it folds smoothly along the scored line. Score with the back of a scissor blade for a softer scored line. Always cut straight edges against a metal ruler or straight edge.

A saw point

1 Before attempting a cut in any size of wood, it is essential to secure the workpiece to prevent it from moving during the cut. When cutting blockboard or plyboard with a crosscut saw, use C-clamps to secure the piece to the bench top. Oiling the teeth of the saw with light machine oil will ease the passage of the saw through manmade boards.

2 Smaller, linear lumber to be cut with a tenon saw can be held in a small vise or hand held on a bench hook. The bench hook rests against the edge of the work bench and has one or two shoulders to support the wood for cutting. When cutting linear lumber on a line, start the cut by making back strokes with the saw, until the cut is established.

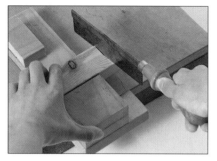

3 Cutting a corner or rounding off an endpiece without the use of a belt sander requires a series of straight cuts that butt onto a curve. Following a template, draw the desired curve onto the workpiece. Using a straightedge, draw a series of cutting lines touching the line of the curve.

4 Cut away the waste with a tenon saw, following the straight lines. Repeat the process with shorter lines, butting the remainder of the curve until the desired outline shape is eventually formed.

5 Using a medium-grit sandpaper wrapped around a sanding block, sand back the cut points to follow the curve. When you are happy with the final line of the curve, finish with fine-grit sandpaper.

6 A miter joint is formed when two 45 degree cut ends make up a right-angled corner. Mark the wood with a knife, using a combination square (see Step 3, p.25), and secure it in a miter box. Make the cut using a tenon or back saw.

7 When using the miter box, and when making a cut in any linear lumber, do not follow all the way through the cut, as the waste will break off and split the edge off the wood. As the cut nears completion, move the workpiece onto scrap wood, clamp it securely, and cut through it carefully, sawing into the scrap. This will result in a clean finished cut.

8 To cut a circle (as required for the CD stack and speaker stands on pp.50–51), scribe directly onto the workpiece using a compass. Protect the center point with tape. Draw consecutive straight cutting lines all around the circle, cut away the waste, and sand to a smooth finish, as in Steps 3, 4 and 5.

Drilling and screwing

1 When drilling a hole for a screw, drill a pilot hole first, using a bit narrower than the screw thread. Drill a hole the same width and depth as the shank and countersink, as shown. Clean the edges of the hole with fine sandpaper, drive the screw with the correct size screwdriver, fill with wood putty, and leave to set. Sand with fine sandpaper before painting.

2 Always use a screwdriver the same size as the screw you are attempting to drive, or you will damage the head. Slotted screws should be lightly greased before insertion so they can drive more easily. Do not attempt more than a third of a turn when using a ratchet driver, as overextending the wrist can result in losing control of the tool. Never use a pump-action or yankee ratchet driver where accuracy and care are necessary.

3 Drilling holes for dowels is simple using an auger bit with a rubber depth stop. Auger bits are self-starting, drilling their own pilot hole with the threaded point, and the spiral thread clears waste rapidly from the hole. To stop the hole, adjust the rubber on the bit to the required depth and drill up to it. To carry on through with the hole, place the work onto a scrap of wood, clamp it securely, and drill the dowel hole straight through to prevent damage as the bit exits the workpiece.

4 The self-starting auger bit requires a ratchet brace to drive it. Control your drilling angle by holding the brace head. Obtain a vertical drive by holding a 90 degree square up against the brace chuck. Drill pilot and shank holes with a twist bit, driving it with a wheel brace. Remember that the twist bit is not self-starting, so punch a small hole first to start the bit in, using a hammer and a center punch, or an awl.

Cutting wood joints

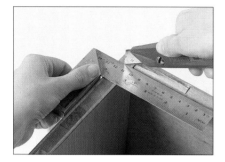

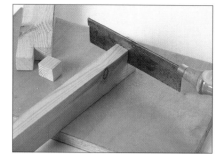

1 When joining boards to form right-angled corners (as with the blanket chest on pp.78–79) use a rabbeted joint. Match the square end to its matching piece, and mark the rabbet across half the depth and then across the entire width of the board.

2 Carefully cut out and remove the waste, using a tenon saw. Work from both sides of the rabbet into the center with the workpiece firmly secured. Check the fit and sand the inside edges of the rabbet. When making a box shape, cut every rabbet before joining the first corner.

3 Simple halving joints (see the fabric screen on pp.100–101) are so-called because they consist of two rabbets, each one half the thickness of the wood. Mark both halves simultaneously with a marking knife, cut and remove the waste.

4 Overlapping halving joints are cut in a similar fashion. The housing is the same width as the smaller dimension of the wood, and half the larger dimension. After cutting the rabbets, check they fit tightly—never attempt to secure a badly fitting joint with an excessive amount of glue.

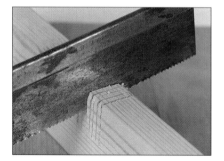

5 To cut a housing for any wood joint, mark the wood with a knife and cut to the required depth. Removing the waste wood from the housing is simplified by making successive parallel cuts with the saw to the same depth.

6 Using a sharp mortise chisel, remove the remaining wafers of waste wood one at a time. Grip the blade, and use the palm of your other hand to drive the chisel through the waste. Never strike the handle end with a hammer; use a wooden mallet. Better still, have the chisel blade sharpened at a hardware store.

7 Simple butt joints or L-joints require accuracy in measuring and positioning. Support and secure both workpieces, making sure they are at right angles to each other, before drilling a pilot hole (to guide the path of the screw), a shank clearance hole (the same bore size as the shank of the screw to half the depth of the pilot hole), and a countersink hole (to set the head of the countersunk screw below the surface of the wood). Finally, drive a screw through the shank clearance hole to take up the inside of the pilot hole.

8 When applying glue to any wood joint, always check the fit first, then spread the glue into the mortise or housing. This prevents the glue from being forced clear on assembly. Remember to use a waterproof adhesive if the joint is likely to be subjected to steam or moisture—in a bathroom or kitchen, for instance.

Nails and punches

1 Drive brads into the workpiece with a pin hammer. Do not damage the surface of the wood by hammering the brad all the way in. Use a nail set or punch to sink the brad below the surface.

2 Drive molding or veneer tacks into edging strip by either drilling a tiny pilot hole first to hold the pin, or by holding the pin steady in a thin cardboard grip. Use a small pin hammer and drive below the surface with a punch as in Step 1.

A new start

A large number of projects can be constructed from secondhand wood. Source unwanted wood from dumpsters and dumps, asking permission first. The belt and tie hanger on pp.74–75 is made from reclaimed wood squared up using its one straight edge, and cut down to size, removing defects, nail and screw holes, and all other signs of its previous life. A country-style effect can be guaranteed by using old floorboards to make the shelf on pp.84–85.

Finished!

Fill any defects in the work with wood putty, then sand and spray or handpaint in the finish of your choice. Light-colored woods can be enhanced by using homemade wood stains. Select your color in an oil-based paint and add a teaspoonful to a jar of mineral spirits, stirring well. Brush onto the wood surface and let it dry. Sand the wood surface back to its original color, leaving your stain effect in the enhanced grain. Finally, brush on a clear protective varnish.

Screwing in

To mount wooden strips or projects to an internal wall, first mark all the positions on the plaster using a level. Drill holes through the strips where needed, making sure the screw will pass through the shank clearance hole cleanly and countersink the holes. Check the strip, using a spike or point to mark the position of the wall holes through the wooden strip. Drill pilot holes first using the spike hole as a starter with a small masonry bit, then move to the correct masonry bit—the size will be marked on the anchor. Clean the hole and tap the anchor into the wall, making sure it fits. Push the screws through the wood so the ends protrude and line them up with the anchor and screw in place. The anchor tightens its grip in the hole as the screw is inserted.

the kitchen

The hub of the home

The kitchen probably needs more planning and organization with regard to storage than any other room in the home, and special care must be taken regarding hygiene and safety. But with careful thought and planning, the kitchen can be a safe and relaxing place to carry out everyday chores.

For many families, the kitchen is the busiest room of the home, a place where members of the household congregate after work or study, to chat about the day's events while a meal is being prepared. For this reason, a pinboard or fridge magnets are often a useful addition in the kitchen so that dates of school events or bills to pay are not cluttering the counter or forgotten. The kitchen is often home to many activities in addition to cooking, such as ironing, or doing homework or crafts at a kitchen table, so thoughtful storage is essential to make the most of this room.

right: A butcher's block will provide a portable additional work surface as well as a versatile storage area underneath, including wine racks and knife blocks.

far right: A large traditional-style hutch with a wooden plate rack and a space for basket shelves underneath is ideal for both displaying and storing.

The "work triangle"

How you arrange your kitchen will depend upon the way it is used. For instance, easy access to cooking utensils and pots and pans is vital to a keen cook feeding a large and hungry family, whereas the reluctant cook will have different requirements. When designing or revamping your work area, keep the sink, fridge, and oven as close together as possible. This is your basic "work triangle" and you will constantly go from one to the other. Make a plan of the kitchen as described on pages 11–12 and draw in the "work triangle". If the total length of the sides of the triangle are less than 5 yards, you will find your work space cramped, and if the total length is more than 6½ yards, you will waste time traveling between the workstations.

Pots and pans in regular use should be kept close to the work triangle. Heavy dishes should be stored in the lower cupboards, and lighter items, such as glassware, on higher shelves. Everyday dishes can be stored further away from the triangle, but within easy reach of the sink or a dishwasher. Open-fronted shelves and hutches are most suited to items that are used and washed often; if they are mainly for display, they are prone to get dusty and greasy, so site them away from the steamy atmosphere of the stove or sink. Add hooks to shelves and cabinets to hang cups and mugs. A wooden plate rack is an attractive means of storage and allows wet dishes to drain.

Extending the work surfaces

A continuous work surface is more versatile than several small areas. Keep the counter as free of clutter as possible, with only utensils and gadgets that are in constant use left out on the work surface. This makes it easier to work and to keep the surface clean. To free up more space, hang cooking utensils on the wall or suspend saucepans and colanders from an overhead wooden rack attached to the ceiling. Sharp knives are safest stored in a knife block or on a magnetic strip attached to a wall.

Traditional butcher's blocks are a stylish and practical way to extend your working surfaces, while at the same time providing extra storage space underneath for vegetables or a wine rack, depending on the individual design. Choose a block with castor feet for extra flexibility.

Simple wire racks are available to attach to the inside of cabinet doors to hang slim utensils and saucepan lids. A shallow wire shelf that clips below the upper shelf in a cupboard is a great way to use up empty space and store small cans and jars. Recycle large plastic ice-cream cartons to hold cleaning materials. Spray paint them different colors for instant identification to what they hold. If stored under the sink, they can be lifted out and carried to the workplace. If floor space is at a premium, hang a small trash can inside the cupboard under the sink. The lid conveniently lifts as the door is opened.

Wall-mounted cupboards should be positioned at least 18 inches above a work surface and be 12 inches in depth at the most. This leaves room for kitchen appliances and storage jars to be kept at the back of the work surface and allows you to work with good visibility without bumping your head. The space above wall cupboards is often wasted. Store a row of wicker shelf baskets above these cupboards, which will make an attractive feature and can hold rarely-used items.

Food storage

Food cupboards must be cool and dry and positioned away from the stove and fridge if possible. Many foods are best kept in cupboards, as they deteriorate in light and warm conditions. Decant bags of rice, sugar, and pasta into square, airtight containers that will hold more than round ones and sit snugly side by side. Choose glass or transparent plastic containers for easy visibility, or label containers well, remembering to add the best-before date. Keep perishable foods that need to have air circulating, such as fruit and vegetables, in baskets or hang garlic and onions from a wall hook.

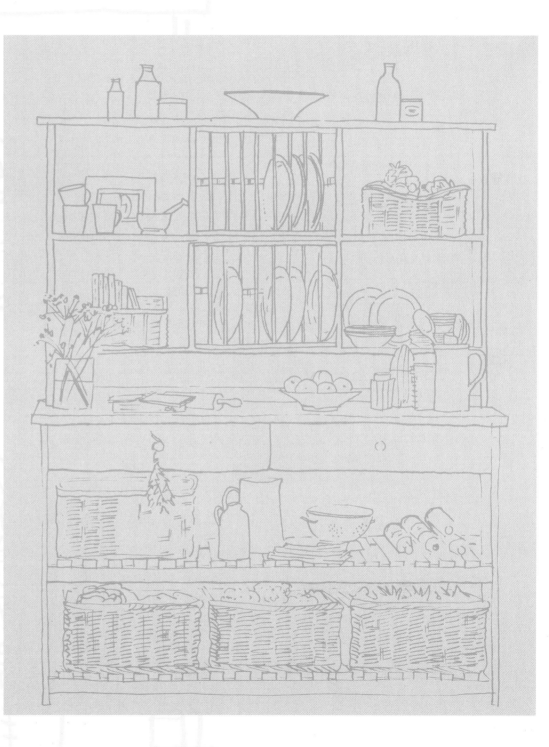

Utensil hanging rack

THIS SIMPLE hanging device will keep essential cooking utensils easily at hand without cluttering the work surface. Attach it to the wall and use traditional butcher's hooks to suspend your cooking implements, which would otherwise take up valuable cupboard space.

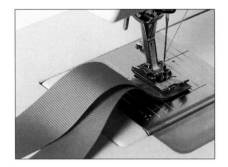

1 Cut two 59-inch lengths of 1-inch wide grosgrain ribbon. Slip two 1-inch D-rings onto each length. Fold the ribbon in half and stitch across the ends taking ⅜-inch seam allowance. Turn right side out and fold in half, adjusting the seam to the back. Slip a D-ring on each end.

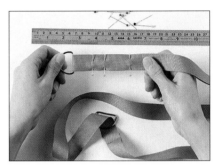

2 Measuring from the top down, mark the following positions in inches across both thicknesses with pins: 1¼, ¾, 1½, 6⅜, 1½, 6⅜, 1½, 6⅜, 1½, ¾. Stitch across the ribbons at the pin positions.

3 Saw four 24-inch lengths of 1-inch wide and ³⁄₁₆-inch thick strips of wood for the slats. Paint the wood with latex paint in a color to match the ribbon.

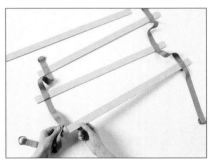

4 When the paint has dried, slip the end of each slat through a 1½-inch wide slot on each of the ribbons.

Tiered vegetable holder

THESE DESIGNER-STYLE mesh bowls have been strung together to make a tiered hanging system to hold vegetables and fruit suspended above the work surface. Be creative and choose colanders, strainers, or baskets for the containers to tie in with the style of your kitchen.

1 Divide the rim of three bowls of graduated sizes into thirds, marking the positions clearly with masking tape.

2 Attach S-hooks to the bowls at the positions marked, anchoring them to the rim or the mesh. Screw a large eye into the ceiling or the underside of a cupboard where you want to suspend the rack.

3 Hang an S-hook onto the screw eye and hang three lengths of chain from the hook. For this holder, 1-yard lengths of chain were used. Hang the hooks on the largest bowl on the ends of the chains.

4 Working up the chains, count the links to attach the hooks of the middle-sized bowl level between the chains. Finally, attach the small bowl.

Wire wall rack

THIS WIRE rack is inexpensive to make, yet has a designer-style country feel, perfect for oil or marinade bottles. The rack has an inner built in frame of twisted wire that is not only attractive to look at, but adds additional strength to the rack to hold the weight of the bottles. Chicken wire is sharp and can be difficult to work with, so always handle it with care, or wear protective gloves.

As an alternative idea, spray the rack in pastel-colored paint and hang in the bathroom to hold decorative shampoo or bubble-bath bottles.

1 Twist ⅟₁₆-inch gauge wire into usable lengths. Once twisted, the wire should be about a third of its original length. Bend a length of wire in half. Clamp the middle of the wire in a vise or looped around a handle. Loop the ends of the wire around either side of a coat hanger. Turn the hanger to twist the wire; the lengths should be about ¾-inch long.

2 Clamp a pencil in a vise. Wrap a length of twisted wire closely around the pencil, 12 inches from one end of the wire, forming a loop. Make another loop 8 inches from the first, in the opposite direction. Bend the wire at a right angle 12 inches from the second loop, using a block of wood. Cut the wire 8 inches from the corner so the wire ends meet, forming a rectangle.

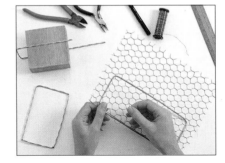

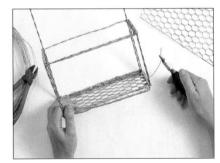

3 Bend four twisted wire frames: one 3 x 8 inches for the top of the holder; one 5 x 8 inches for the front; and two 3 x 5 inches for the sides. Make a base frame from ⅛-inch gauge wire for extra strength. Cut a piece of chicken wire 7 x 9 inches. Fold in half lengthwise over one long edge of the base frame.

4 Bend the cut edges of the chicken wire over the frame. Use thin wire and a pair of pliers to lace the frames together edge to edge, forming the box shape, attaching the corners securely. Attach the base frame by lacing it with ⅟₂₄-inch gauge wire for extra strength.

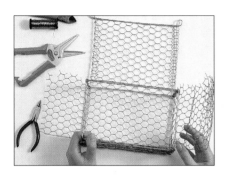

5 Place the twisted wire skeleton with front uppermost on a length of 12-inch wide chicken wire. Wrap the chicken wire over the right, front, and left side of the rack, cutting away the waste as you work. Secure the ends of the chicken wire to the skeleton and use thin wire to secure at the corners.

6 Make a loop at the end of a length of twisted wire as in Step 2, then make another loop 8 inches from the first. Position across the back at the top of the "box" with the loops pointing outward. Bind securely to the rack with thin wire. Bend a length of twisted wire into a heart shape and attach to the front with thin wire.

Flatware tray

ORGANIZE SILVERWARE in this functional felt-lined box designed to help keep it neat and free from scratches. This versatile tray could also be used in an office drawer to sort pens, paperclips, and other paraphernalia, or to store a selection of small toys in a child's bedroom.

The tray can be made from either ³⁄₁₆-inch thick foam board (which is available at art-supply stores) or pieces of corrugated cardboard. Use a sharp craft knife when cutting felt and replace the blades regularly, as it can make them become blunt very quickly.

1 Cut two rectangles from foam board or corrugated cardboard 2 x 10¼ inches for the front divisions and 2 x 12¼ inches for the back division. Spread craft glue evenly and sparingly on one side. Press to the felt. Cut the glued felt even with one long and both short edges of the boards. Divide the back divisions widthwise into thirds with a pencil.

2 Spread glue on the narrow upper edge and the other side of the front divisions. Press the felt smoothly on top and cut away the excess. Fold the felt on the back division over the board and cut it level with the edges of the board. Glue the ends of the front divisions against the pencil lines on the back division. Leave to dry.

3 Cut the felt above the front divisions to within ¼ inch of the back division. Spread glue on the narrow upper edge and inner side of the back division. Press the felt on top. Carefully trim away the excess felt around the joints.

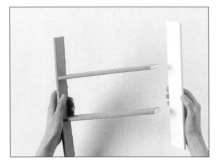

4 From foam board or corrugated cardboard, cut two rectangles for a box front and back 2⅜ x 12¼ inches and two rectangles for the box sides 2⅜ x 14½ inches. Use a pencil to divide the front widthwise into thirds. With the lower edges even, glue the ends of the front divisions against the pencil lines on the front.

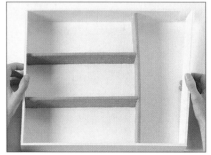

5 Glue the ends of the box front, back division, and box back between the box sides, forming the box shape. Set aside, leaving the glue to dry completely.

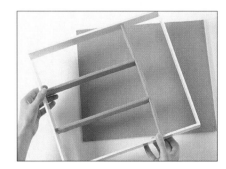

6 Draw around the construction on foam board or corrugated cardboard and cut out the base. Glue felt to one side of the base. Glue the box onto the base and weight it with books while the glue dries.

7 Cut a strip of felt 6 x 55 inches to cover the rest of the box. Starting at a back corner and with ¾ inch below the base, glue the felt to the outside of the box, overlapping the ends.

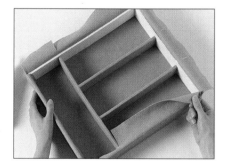

8 Cut the felt to the corners and to ⅝ inch from the box at the divisions. Glue the felt under the base and inside the box. Carefully trim the felt at the edges with a craft knife. Cut a rectangle of felt ⅜ inch smaller than the base and glue it firmly underneath.

Hanging bag organizer

THIS NIFTY device is great for storing all sorts of kitchenware that you want to keep close at hand, and it can be attached to a wall or the back of a door. The two pouch bags have stiffened tops to hold them open for easy access, and the bag dispenser is designed so you can push recycled plastic bags down through the top and pull a bag out from the bottom when you want one.

These pouch bags and dispenser have been made from cheap but durable waffle-cotton dishcloths. Two additional hooks attached to the dowel provide a handy place to hang a couple of matching cloths. Use a ⅜-inch seam allowance for the pouch bags and the dispenser.

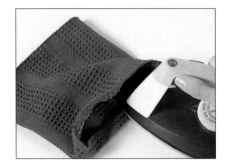

1 Cut fabric 7 x 15 inches for a bag. With right sides facing, stitch the short edges together. Press under ⅜ inch, then ¾ inch on the upper edge. Stitch close to the lower pressed edge to make a channel, leaving a 1½-inch gap at the seam.

2 Refer to the template on p.130 to cut a base from fabric. With right sides facing, pin and stitch to the lower edge of the bag, matching the dot to the seam. Turn the bag right side out.

3 Cut a 17-inch length of ½-inch wide plastic boning. Cut one end to a point and stick a piece of tape over it to help it slip through the bag channel. Thread the boning through the channel. Stitch the gap closed. Make another bag the same way.

4 Cut the dispenser from fabric 12 x 13 inches. Pin ½-inch wide bias binding 1¾ inches from both long edges. Stitch close to the pressed edges. Thread 8 inches of elastic through each channel. Pin in place. With right sides together, stitch the short ends. Press open and finish the raw edges. Press ¼ inch then ⅜ inch to the inside on the upper and lower edges. Stitch in place.

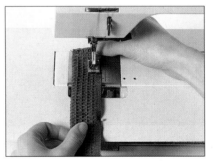

5 Cut a strip of fabric 3 x 7½ inches for a strap for each bag and dispenser. Fold lengthwise in half with right sides together. Stitch the raw edges, leaving a gap to turn. Clip the corners and turn right side out. Slipstitch the opening closed. Stitch a piece of touch-and-close tape to one end of the strap and a corresponding piece to the other end.

6 Pin then stitch the strap to the back of each bag and dispenser. Screw a cup hook 1½ inches from each end of a 21-inch length of 1-inch diameter wooden dowel. Mount a 1-inch metal end bracket to the wall, insert one end of the dowel, slip an end bracket on the other end and mount it on the wall. Slip the bag and dispenser straps over the dowel and fasten the straps.

Bottle bag

HERE IS a softly padded canvas bag with four corner sections to hold bottles snugly and safely. The bottle sections are separated from the middle storage area of the bag by deep pockets into which you can slip small freezer packs in order to use the bag as a cool bag for food shopping or picnics. As the bag is padded, it is ideal for carrying other delicate items, such as photographic equipment.

Choose a firm, hardwearing fabric like the canvas which has been used here. Start by cutting two bottle divisions 4½ x 14¼ inches from fabric. Fold widthwise in half with right sides together and stitch the short edges. Use a ⅜-inch seam allowance in Steps 1–6.

1 Turn right side out. Cut four outer pockets 4½ x 8¾ inches. With right sides together, pin each bottle division to an outer pocket with the seam ⅜ inch above the lower short edge. Pin the other outer pocket on top and stitch the pinned edges. Cut two outer pocket linings and two inner pockets from fabric 8 x 8¾ inches.

2 Baste the outer pocket linings to the wrong side of the outer pockets and treat them as one. Press under ⅜ inch on the upper short edges of the outer and inner pockets. Pin and stitch a set of touch-and-close strips on the pressed edge of the inner and outer pockets. With right sides together, stitch the inner to the outer pockets on the short lower edge.

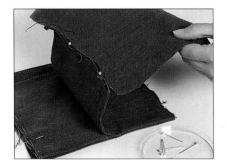

3 Turn the pockets right side out and pin raw edges together. Cut four bottle sections 7 x 9¼ inches. With right sides together, pin the raw edge of each bottle division to one long edge of a bottle section ¼ inch above the lower edge of the bottle section. Pin the other bottle section on top and stitch. Press the seams open.

4 Pin the bottle sections to the pockets with the pockets' lower edge ¼ inch above the lower edge of the bottle sections. Cut two middle sections 4½ x 9¼ inches and with right sides together, pin and stitch one long edge to the pinned edges of one pocket, matching the lower edges of the bottle and middle sections. Stitch. Repeat on the other pocket.

5 Cut two rectangles of fabric and lightweight batting 9¼ x 17 inches for the outer bags. Cut two 33½-inch lengths of webbing for the handles. Pin the ends of each handle to the fabric outer bags 4½ inches in from the short edges. Stitch close to both edges of the webbing for 7½ inches and stitch a cross at the top of the stitching to reinforce.

6 Baste the batting under the fabric outer bags. With right sides together, stitch the short edges. Trim the batting in the seam allowances; press the seams open. With right sides together, pin and stitch the outer to the inner bag at the upper edges. Trim the batting in all seam allowances. Press the outer bag to the outside. Baste both bags together at the lower raw edges.

7 Refer to the diagram on p.130 to cut two bases from fabric. Also cut one base from corrugated cardboard, cutting it ⅝ inch smaller on all edges. Place the cardboard base centrally between the fabric bases. Pin and baste the fabric bases together, enclosing the cardboard base. With wrong sides together, pin and baste the base to the lower edge of the bag, matching raw edges.

8 Cut a bias strip of fabric 2¼ x 34½ inches. Press lengthwise in half to make a binding. Turn under one end to start, then pin and stitch the binding to the lower edge of the bag, making ¼-inch seam allowance. Turn the binding to the underside and baste then stitch in place.

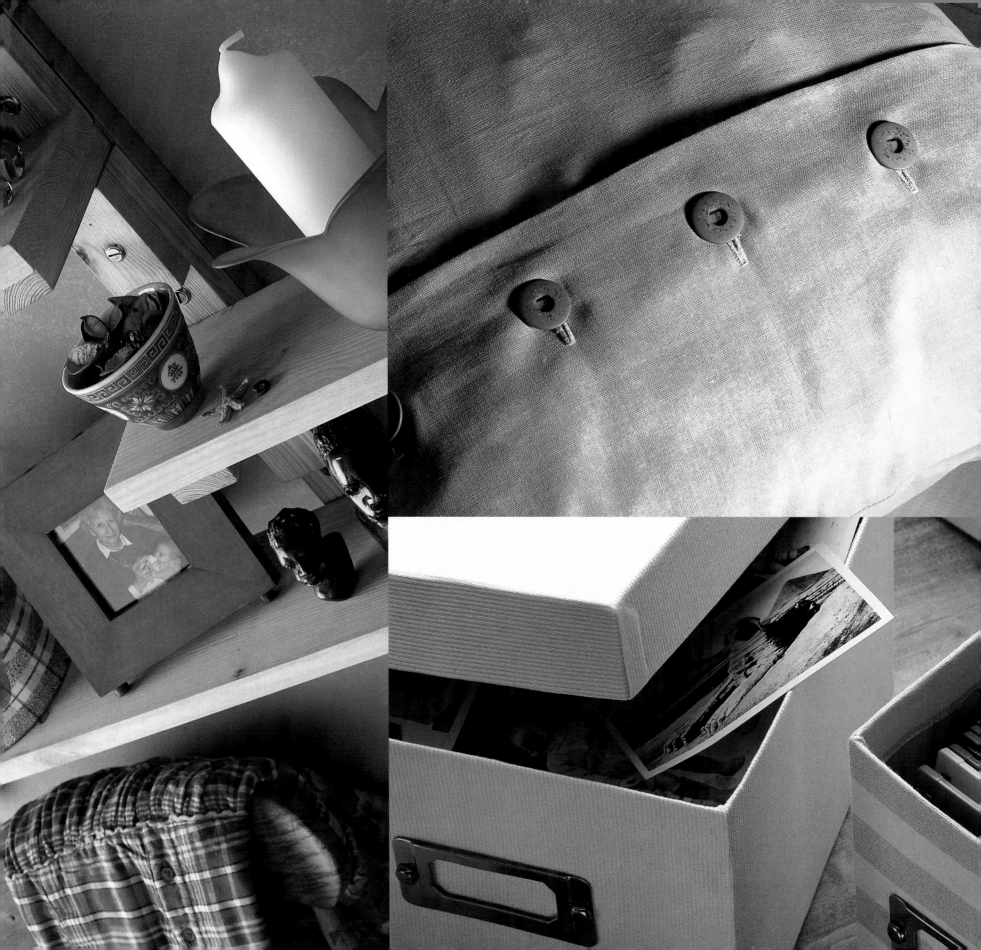

the living room

The comfort zone

below: **A living room designed so that the storage units and surfaces are an integral part of the seating area, providing a place for lamps and plants, and making the most of a small space.**

The living room is a place where you retreat to at the end of the day to relax, be yourself, and put your feet up. But it is also the public face of your home, as the main room to which visitors are invited, so any storage ideas must be stylish and blend in with the decor. The living room has many personalities according to the variety of activities that it hosts, from a quiet reading room, to a children's playroom, to a party venue complete with music—so it must cater for many needs.

If you are planning the room from scratch, decide on your storage needs before choosing fabrics and color schemes, then you can subtly disguise some items by using matching colors. Alternatively, as you will probably spend a lot of time in the living room, be creative and make imaginative items that will be fun to live with.

Hide or display

Storage usually falls into two categories in the living room—for display or concealment. Treasured collections of glass or porcelain are best displayed in glass-fronted cabinets to minimize dust, or on shelves out of the reach of young grabbing hands or boisterous pets. Choice of lighting needs careful thought to enhance the pieces. Paint the inside of a display alcove white to reflect light; a strong color may fight the display for attention. A collection of books can be very heavy, and the shelves they sit on must be well supported against structurally sound walls. Allow at least an inch between the top of the books and the next shelf. Alcoves always make an excellent space for shelving, and a cabinet incorporated underneath will be accessible for children to store their toys.

If you want to bring a bit of elegance to your living room, it is worth looking at traditional-style pieces of furniture, which are often well-equipped for maximum storage. An antique armoire—a large French-style clothes closet—would look as good in a living room as it would in a bedroom and can have shelves added to store drinks, glassware, and games. A classical bureau has always been a popular piece of furniture, and this is not surprising given that with a fold-down or roll-up lid, it becomes an instant desk with many small handy compartments for stationery, as well as having large drawer space underneath.

Attractive pieces of furniture can be adapted for storage. A old wooden trunk used as a coffee table provides ample storage for seldom-used items—perhaps for bedding for overnight guests if you have a sofa-bed in the living room. Alternatively, place it against a wall or under a window with cushions on top,

so it can provide extra seating for young children. It is also a great place to quickly hide children's toys and other untidy items when guests arrive.

If you are buying side tables or a coffee table, opt for a style that has drawers or a shelf underneath for books and magazines. This allows you to have a quick clear-up if visitors drop in unexpectedly. Use stylish accessories such as baskets and boxes that complement the room's decor to hold needlework or other hobby materials, so that they enhance rather than detract from the room.

Dining areas that share the living room can be separated with a room divider that provides storage in the form of shelves or cupboards on one or both sides. A screen makes a quick, improvised room divider and can be folded away when not needed, creating privacy when different activities are going on.

Living with technology

In this age of technology, you can expect to find a television, video, and stereo system in nearly every living room. As much as we may enjoy the entertainment that these gadgets provide, with their black and gray exteriors, they do little to enhance the decor of the living room.

The television is often the focal point in a room and also has its own special requirements. It must be positioned away from uncovered speakers, steel or cast iron objects, and away from direct heat or where sunlight will fall across the screen. Televisions and sound systems on shelves need space for air to circulate and to accommodate wiring.

If you want to minimize the impact of the television, consider a swivel TV arm that will support the weight of a TV and can be maneuvered 180 degrees to swing the TV out of sight when it is not being watched. A TV and video cabinet is a stylish way of hiding these items completely from view—it also has the added bonus of possibly discouraging too much TV watching! Invest in small, wall-mounted speakers to keep more

floor space free or create dual-purpose speaker stands which incorporate a stash of CDs (see pages 50–51).

Unsightly electrical cables can be slipped into plastic sleeves or run under the floor out of the way. Consider running a sound system from a multisocket extension that will organize the mass of wires from electical outlets. The TV and video can be run in this way, too. This will create an attractive, neat and safe environment in which you can relax.

above: Conceal a video recorder and tapes in a purpose-built cabinet, choosing a style of unit to coordinate with the furniture in the room.

CD stack and speaker stands

STACK YOUR favorite CDs in this dual-purpose speaker column stand. The slim shape means that this storage item can fit unobtrusively into any small corner of the living room. If you want a streamlined finish, spray-paint your stand to match the color of your speaker cabinets.

To successfully create the circular top and base, refer to Step 8 on page 26 in Techniques.

1 Cut two side pieces 5 x 24 inches, one back piece 6½ x 24 inches, two insert end pieces 5 x 5 inches and two circles of 9-inch diameter from ¾-inch thick blockboard. Draw the positions of the two insert end pieces on the circles, following the template on p.130.

2 Cut edging strip to 5-inch lengths for support pieces. Pin to the side pieces parallel to the base. The number of strips can be varied to suit your CD collection. Make sure that both side pieces are pinned identically.

3 Screw the back on flush to the sides, screwing through the back with 1½-inch countersunk screws. Countersink the heads enough that they can be filled with wood putty prior to spraying.

4 Screw the circular top and base to the insert pieces with 1¼-inch countersunk screws. Screw into position at the ends of the stand, securing at the center of the sides and back with 1¼-inch countersunk screws. Fill all countersunk holes and any wood defects with wood putty. Sand the unit with fine sandpaper and spray the interior and exterior.

Magazine rack

PILES OF newspapers, magazines, and catalogs scattered around the living room plague many a household and can quickly become messy. Keep them neatly stacked, but easily at hand, in this mixed-media finished stand, which can be tucked away to the side of a sofa or armchair. The angled cutouts make access easy and give the piece a distinctive modern look.

Refer to Techniques, pages 28 and 26, for advice on cutting rabbets and creating the rounded ends.

1 Cut four 20-inch lengths of 1 x 2-inch wood for the leg assembly. Halfway along the lengths, mark a half side rabbet ¾ x ¾ inch to overlap the joints. Cut the rabbets and chisel out the waste.

2 Slot together the overlapped leg assemblies in pairs. Mark the two ends of each pair that will be the feet 8½ inches from the rabbet, then mark one upper end 4½ inches and the other upper end 9½ inches from the rabbet respectively. Cut the feet at 45-degree angles to form a base. Round the other ends and sand the cut edges.

3 Glue and join both leg assemblies. Clamp together with a C-clamp, wipe off any excess glue with a damp cloth, and set aside to dry.

4 Cut two pieces 10 x 12 inches and 6 x 12 inches from ¾-inch blockboard. Cut a 5 x 12-inch and 2½ x 12-inch angled recess in the boards, using the diagram on p.131. Join the pieces together, gluing and screwing with 1¼-inch countersunk screws. Countersink the heads enough that they can be filled and sanded.

5 When the glue has set, drill a hole in the leg assemblies 2 inches from the shorter rounded end and 4 inches from the longer rounded end to accommodate a 1¼-inch countersunk screw. Varnish the leg assemblies with clear varnish and spray-paint the holder. Leave to dry.

6 Screw the magazine holder to the leg assemblies through the previously drilled holes with 1¼-inch countersunk screws and brass seatings.

Adjustable shelving system

THE VALUE of shelving should never be underestimated as a cheap and essential item of storage in most homes, whether for purely practical purposes or for display. This clever geometric system is all cut from the same lumber stock, and the horizontals can be slotted into any or all of the nine equal spacings, giving an interchangeable support system. The finished effect is a strong, flexible system, with the wall fixtures for the uprights neatly hidden behind the supports. The measurements given here can, of course, be adapted to suit your own requirements.

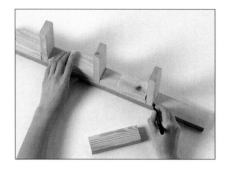

1 Cut three uprights 24 inches long from 1 x 2-inch wood. For supports, cut 21 pieces 5 inches long and nine pieces 3 inches long from 1 x 2-inch board. Position the pieces following the diagram on p.131 and mark them on the uprights.

Attach the first layer of supports to the upright using the markings as a guide, by center drilling 1½ inches from the bottom of the support and screwing through to the upright using 1¼-inch countersunk screws and brass seatings.

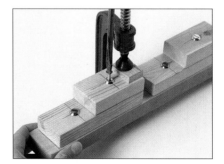

3 Attach the nine second-layer supports under the position of the horizontal supports that will hold the shelves, by center drilling 1½ inches from the bottom of the support and screwing through to the first layer support with 1¼-inch countersunk screws and brass seatings.

4 Position the horizontals that will hold the shelves into the slots, resting on the six second-layer supports. Make sure that they are a reasonably tight fit. Position the shelves onto the horizontals. These shelves are 5 inches wide. When you are happy with the arrangement, cut off any waste over ½ inch long at the top of the upright, remove the shelves, and sand and varnish all the pieces.

Video box

STORE AND select a video at a glance with this strong stylish box that holds up to eight tapes. Create a set of these video boxes that can be stored freestanding on shelving, with each one holding a different viewing theme, for instance, movies, cartoons, or sport.

Customize the box by covering it in a fabric to coordinate with your decor. Choose a matching colored mat board that makes an attractive interior that will be visible when the box is not full.

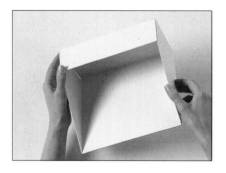

1 Refer to the diagram on p.131 to cut a video box from mat board. Score along the broken lines (if you are using colored mat board, score the white side). Fold back along the scored lines, bringing the lower edge of the sides to meet the edges of the base. Fold the front up.

2 Cut 1-inch wide strips of white paper. Fold lengthwise in half. Glue the strips along the seams with paper glue. Cut the strips even with the box.

3 Cut a rectangle of fabric 9 x 19 inches and apply spray adhesive to the wrong side. With the wrong side of the fabric uppermost, press the box front centrally on top with approximately ¾ inch of fabric below the lower edge. Glue the fabric smoothly to the sides.

4 Glue the excess fabric to the back of the box and under the base, trimming and folding the fabric under neatly at the corners. Cut the fabric even with the diagonal edges of the sides, upper front, and back edges. Cut a rectangle of paper 8¼ x 12 inches. Glue to the back and base of the box with spray adhesive, covering the raw fabric edges.

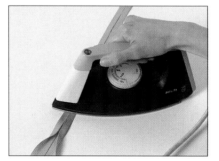

5 Cut a bias strip of coordinating fabric 1½ x 32 inches to bind the edges. Draw a line along the center of the binding with an air-soluble pen. Press the edges to the center. Set the binding aside until the line disappears.

6 Use all-purpose household glue sparingly to stick the binding over the edges, folding under the fullness at the corners and turning under the end to finish. Cut off the excess.

Guest comforter bag

SPARE BEDDING used by occasional guests can take up a lot of storage space, considering that it is probably not used very often. If you store a spare comforter in an attractive bag, it can double as a big, squashy cushion in the living room and give you extra seating, too. This cushion bag will hold a double-bed quilt. Fold it in half, then roll it to push it inside the bag.

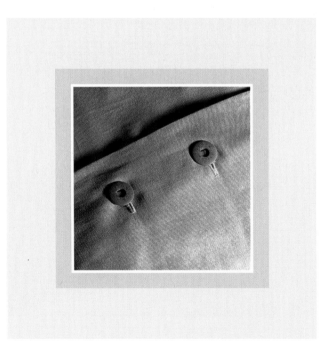

1 Cut a rectangle of fabric for the flap 17 x 24 inches. Press lengthwise in half with wrong sides together. Make five equidistant buttonholes 1½ inches from the pressed edge.

2 From contrasting fabric, cut two rectangles for the back and front 24 x 42 inches. With right sides together, pin the flap to one short edge of the back, matching raw edges.

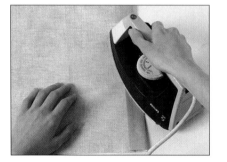

3 Press under ⅜ inch then 1¼ inches on one short edge of the front. Stitch close to the first pressed edge to hem it.

4 With right sides together, pin the front and back together, matching the raw edges and with the front hemmed edge on top of the flap. Stitch the outer edges, making a ⅝-inch seam allowance. Clip the corners and turn right side out with the flap over the front. Sew on buttons to match the buttonholes.

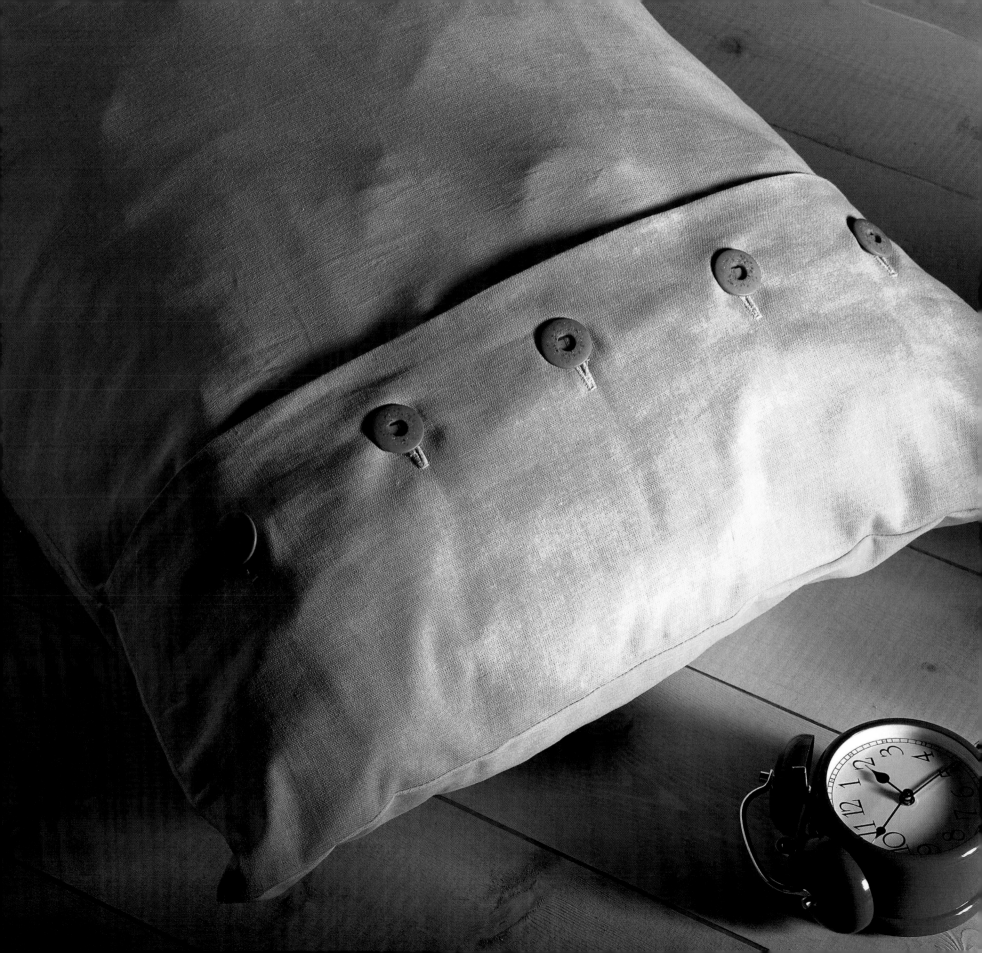

Storage boxes

THESE POPULAR storage boxes with matching lids and metal labels are quickly becoming an essential but fun item to have in the home. Stylish fabric-covered boxes are often expensive to buy, but they can be easily crafted from fabric scraps and thick cardboard.

Choose a mediumweight fabric, and select colors and a design that will match the furnishings in your room. This set is custom-designed to hold three popular items: cassettes, photographs, and floppy disks.

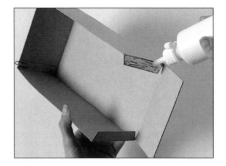

1 Refer to the diagrams on p.132 to cut a box and lid from thick cardboard. Score with a craft knife and ruler, then fold back along the broken lines. Glue the box tabs under the sides with craft glue. Hold with paperclips while the glue dries.

2 To cover the box, draw a rectangle on fabric 5 x 31 inches for the cassette box, 6½ x 27 inches for the photo box, and 6½ x 23 inches for the floppy disk box. Spread glue along the outline on one long edge to prevent fraying. Leave to dry, then cut out. Spread glue evenly and sparingly on one side of the box. Press the fabric on top with ⅝ inch extending beyond one end and the stiffened edge extending ¾ inch above the box.

3 Smooth the fabric out from the center and glue down the extending end.

4 Continue gluing the fabric around the box, turning under the raw end and sticking it in place. Glue the upper edge inside the box.

5 Trim the fabric to the corners of the base and glue under the base, cutting away the fabric at the corners so it lies flat. Cut a rectangle or square of paper ⅜ inch smaller than the base and glue it centrally under the base.

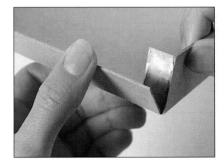

6 Cut 1-inch wide strips of paper and fold in half lengthwise. Bring the ends of the lid sides together and glue the strips inside the seams. Hold in place with paperclips while the glue dries.

7 Spread glue evenly and sparingly on top of the lid and place face down on the wrong side of the lid fabric. Smooth the fabric out. Draw around the lid 2 inches from the edges. Spread glue sparingly on the outline to prevent fraying. Leave to dry, then cut out. Cut from the fabric corners to the lid corners. Trim the corner cuts to ⅜ inch beyond the lid edges.

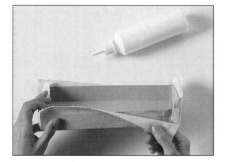

8 Glue two opposite edges of fabric to the sides of the lid, sticking the extending ends around the corners. Glue the fabric to the remaining lid sides, gluing under the excess fabric at the corners. Glue the fabric inside the lid. Attach a metal label to the box.

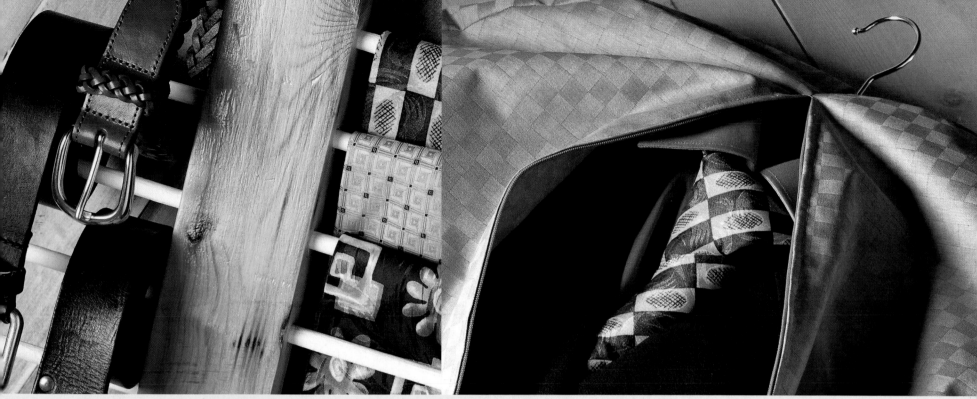

the bedroom

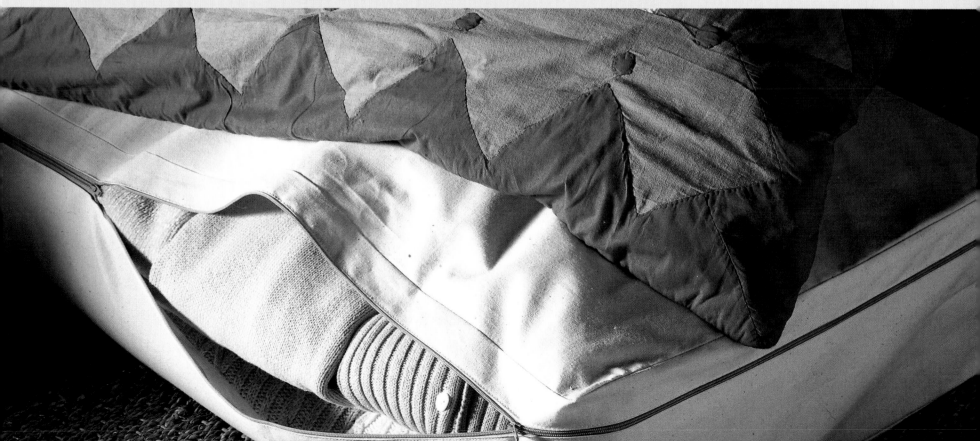

A haven of dreams

Most of the hours spent in the bedroom we are asleep, so we are completely unaware of our surroundings. But to induce this state of relaxation, the bedroom should be a peaceful, tranquil environment to encourage rest, and a clutter-free room is the first step.

Clothing and bedding are the main items that need to find storage space in the bedroom, and they have similar requirements. Get out of the habit of shedding worn clothes on a chair, or worse still, the floor! Put clothes away—they will last longer and look better—or invest in a wooden clothes valet or antique dressmaker's dummy to keep a regularly worn jacket or other items on hand.

Storage starts with the bed, which often takes up the majority of space in the room. If the bed has legs, use the space under the bed to keep storage items such as suitcases or a zippered fabric box (see pages 66–67). A dust ruffle around the bed base will hide and protect the containers underneath. When buying a new bed, consider one with storage built-in underneath, either with drawers to pull out or with a cantilevered base that lifts to reveal an area the size of the bed.

Hanging or storing clothes

Apart from the bed, the closet will be the other main focus in the room, and because of the amount of space it will occupy, it is worth considering something a bit more stylish than white louvered doors and mirrors. Old armoires can be picked up inexpensively from secondhand stores and a decorative paint treatment will soon give it a new lease on life. Another contemporary option is a canvas-covered closet, or for a cost-saving idea use a freestanding clothes rod that can be wheeled about on castors.

If you are creating a built-in closet, remember that an adult's closet should be at least 24 inches deep. A row of built-in closets can make the most of an awkward space and installing bifold or sliding doors means that the closet won't intrude into the room.

If you have a closet positioned within an alcove, hang a curtain in front your clothes for neatness and protection from sunlight and dust.

A closet's contents need disciplined editing; be ruthless in getting rid of clothes that you don't wear. Clothes keep their shape better if they are hung on wooden hangers instead of wire

right: **Instead of throwing clothes onto a chair in a crumpled heap, keep them neat and free of creases by hanging them on a wooden valet.**

far right: **A modern-style closet, which consists of a portable clothes rod with a set of canvas hanging shelves and a zippered closet bag, makes the most of the space in a small bedroom.**

ones. Hang clothes of the same length together—long at one end and short at the other. This means that a small set of drawers can stand inside the closet under shirts, blouses, and jackets. Hooks or wooden strips inside the doors can hang belts, ties, and scarves. Keep everyday shoes in a fabric shoe rack to hang inside a closet door or stacked on a rack.

Out-of-season clothes are best stored elsewhere in bags or boxes with cedar balls or blocks (the aroma of cedar deters moths). If your cedar-wood accessories have lost their fragrance, sand them lightly or rub a little cedar oil into the wood.

Small accessories can easily go astray. Roll socks in pairs and clip pairs of gloves together. Do not leave jewelry lying around loose, as it can easily get lost or damaged and should also be kept safe for security purposes. A soft jewelry roll can be kept in a drawer and also used when traveling.

Creative storage ideas

Traditional items often make original and distinctive storage containers. Store bed linen and blankets in a steamer trunk that will double as a seat with a few soft cushions thrown on top. Old-fashioned leather suitcases or hat boxes can look rather grand stacked on top of an armoire and will hold a lot of occasionally-used items.

In a small living space, the bedroom may also have to be a used as a home office or mini-gym. This area can be disguised with a screen in the corner of the room that can hide daytime activities. Decorate it to complement the decor on the side facing the room and use as a memoboard on the reverse side.

Make a small, plain circular table into an elegant dressing table by slipping a floor-length piece of fabric over it. Have a glass top with beveled edges made to fit the surface area. This will make it easy to keep the surface clean. Display a lamp, mirror, perfume, and ornaments on the top and use the space underneath to hide things you'd rather keep out of sight, such as boxes or pairs of shoes.

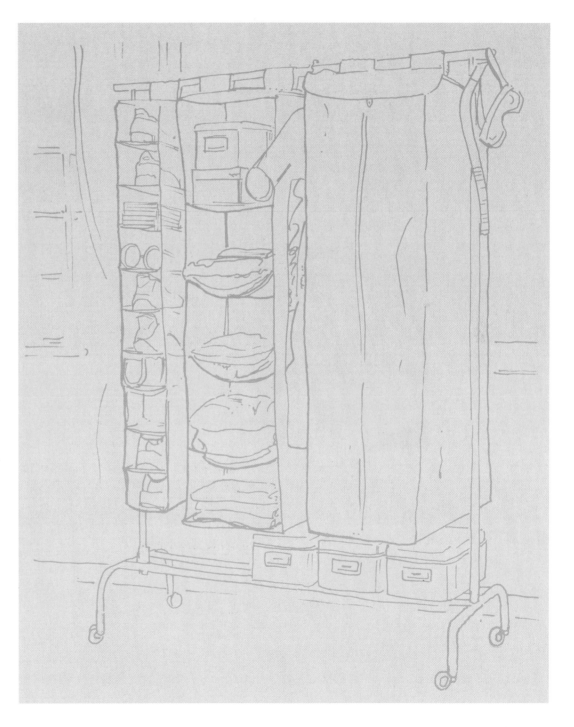

Underbed box

THIS SLIM fabric box can be hidden under the bed to make practical use of this dead space. Use it to store spare blankets and bed linen or out-of-season clothing. The zipped cover keeps the contents secure and free from dust. Slip a block of cedar wood inside to keep away the moths.

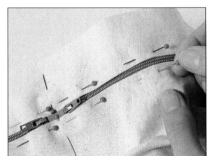

1 Cut a strip of heavyweight canvas 5¾ x 78¼ inches for the base band and a strip 2¼ x 78¼ inches for the lid band. Cut a strip 7 x 42¼ inches for the back band, and cut two rectangles 18½ x 42¼ inches for the lid and base. Press under ⅜ inch on one long edge of both bands. Mark the center of the pressed edges with a pin.

2 Pin two 39-inch zippers under the pressed edges with the top ends meeting ⅜ inch each side of the center pins, positioning the pressed edges against the zipper teeth. If you are using zippers that have been cut from a continuous length, they will be unfinished at the top and the zippers could come off the end, so pin the teeth at the top of the zipper under the seam allowance. Stitch close to each side of the zippers.

3 With right sides together, stitch the ends of the zippered bands and back band together, starting and finishing the seams ⅜ inch from the ends. Press the seams toward the back band. Slightly unzip the zippers so the box can be turned right side out.

4 Open out the seam allowance at the top of the lid band. With right sides together, pin the back band to one long edge of the lid. Pin the lid band to the remaining edges of the lid, clipping the band at the front corners of the lid. Stitch, making a ⅜-inch seam allowance. Stitch the base and lower band the same way.

Dress cover

A DRESS cover will keep clothes dustfree and protect them from getting snagged on other items. This long-length cover will hold a full-length outfit, so it would be ideal to hold a precious wedding or evening dress. A fabric loop at the bottom can be hung over the coat hanger hook for carrying, making this a vital item for traveling and vacations, as well as for storing seldom-worn items in the closet.

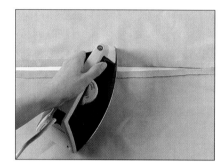

1 Lengthen the dress cover pattern on p.133 by 48 inches to cut two fronts along the front cutting line and one back to the fold, using lightweight fabric. With right sides together, pin the fronts. Stitch the center seam for ¾ inch at the upper end and 7 inches at the lower end making a ⅝-inch seam allowance. Press the center seam open.

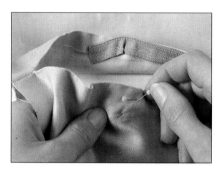

2 With the base of the zipper at the bottom of the opening, pin, then baste a 43-inch zipper with the pressed edges meeting along the center of the zipper. If you are using a zipper that has been cut from a continuous length, it will be unfinished at the top, so pin the teeth at the top of each side under the seam allowance. Use a zipper foot to stitch ¼ inch each side of the center, continuing the stitching to the upper edge of the fronts.

3 Cut a strip of fabric 2¼ x 6¼ inches. Press under the long edges by ⅜ inch then press the strip in half lengthwise. Stitch close to both pressed edges. Pin and baste the ends 1¼ inches from each side of the center front seam on the lower edge.

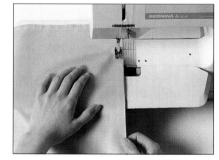

4 Pin the front and back with right sides together. Stitch the outer edges, making a ⅜-inch seam allowance and clip the curves and corners. Turn right side out and press. Topstitch ⅜ inch from the outer edges.

Suit cover

THIS SIMPLE-TO-MAKE suit cover will hold a suit or a number of lighter garments, either for storage in the closet or for traveling. Based on the design of the dress cover on pages 68–69, it is shorter and made with an additional gusset to accommodate bulkier garments. It has been made from sturdy plastic-coated fabric to provide extra protection for your clothes from the effects of dust or the weather.

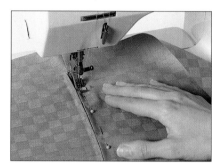

1 Lengthen the suit cover pattern on p.133 by 32 inches to cut two fronts along the front cutting line and one back to the fold, using plastic-coated fabric. With right sides together, stitch the center front seam for 1½ inches at each end, making a ⅜-inch seam allowance. Press the seam and center front edges open.

2 Pin a 38-inch zipper under the open edges with the pressed edges meeting along the center of the zipper. Use a zipper foot to stitch across the ends and ¼ inch each side of the center of the zipper.

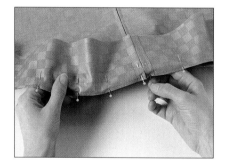

3 Cut two strips of fabric 3¼ x 60 inches. With right sides together, stitch the ends, making a ⅜-inch seam allowance to make one long strip. Press the seam open. Press under ⅜ inch on the ends. Stitch in place. With right sides together and the ends of the strip meeting edge to edge at the center front seam, pin the strip to the front, matching the seams.

4 Clip the strip at the corners of the cover so it lies smooth. Stitch, making a ⅜-inch seam allowance. Stitch a few times over the top of the center front seam to reinforce. Stitch the back cover to the strip the same way. Turn right side out and press.

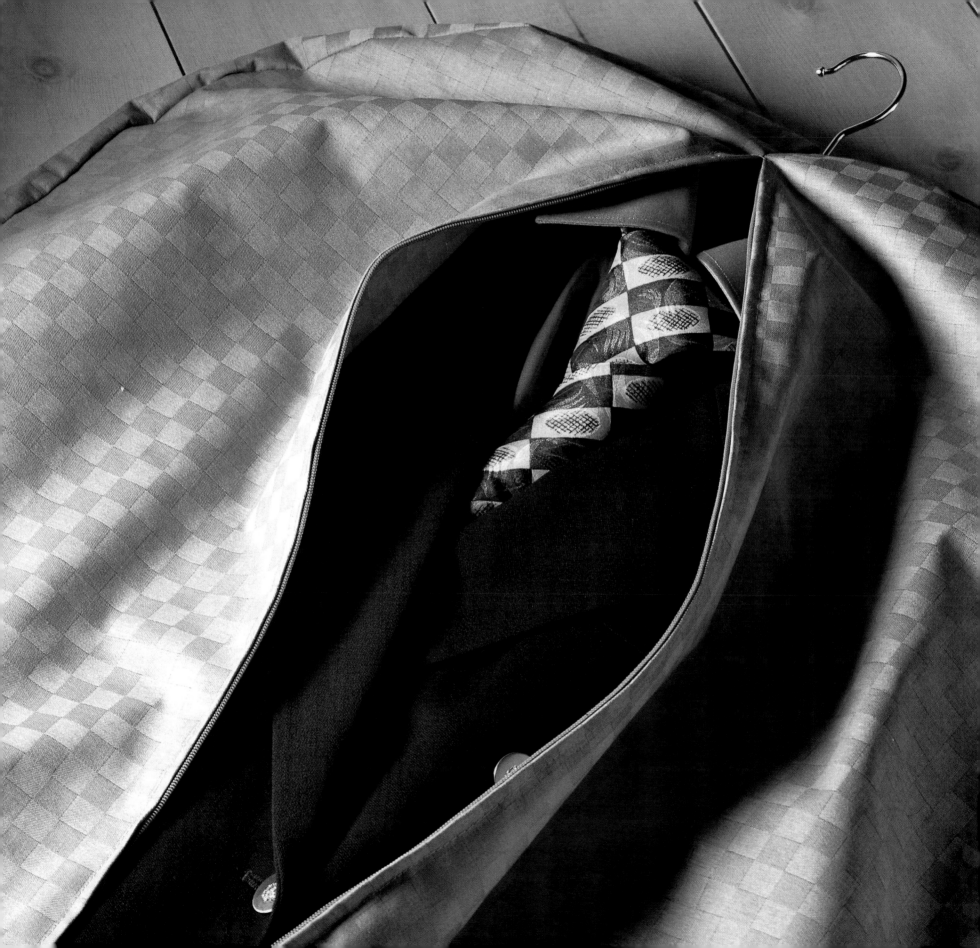

Hanging sweater shelves

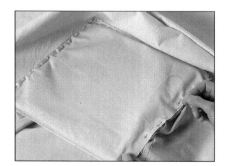

A SET of stylish canvas shelves is a very practical way of storing knitwear or T-shirts in a closet, especially if you have limited drawer space. The shelves are attached using the touch-and-close detachable hanging strap at the top, which fastens neatly and securely over a clothes rod. This system has the added bonus of being portable—the lightweight canvas can be folded flat and easily transported in a suitcase for instant and convenient storage when you travel.

To start the hanging sweater shelves, refer to the diagram on page 133 to cut one support from canvas, and stitch a 1-inch hem on the short edges. Make a ⅜-inch seam allowance in Steps 6 and 7.

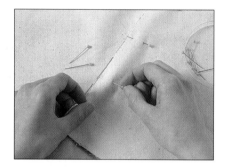

1 Cut three canvas shelves 12¼ x 24½ inches. Press widthwise in half. Topstitch ¼ inch from the pressed edges and pin the opposite raw edges together. Draw the broken lines on the wrong side of the support with an air-soluble pen. Pin the shelves centrally to the support, matching the pinned edges to the broken lines.

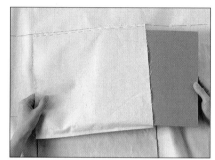

2 Tack ¼ inch from the pinned edges, starting and finishing ¼ inch inside the side edges of the shelves. Cut five 10¾-inch squares of corrugated cardboard. Slip a square centrally inside the middle shelf. Pin the raw side edges together enclosing the square.

3 Fold the support around each side of the middle shelf, matching the raw edges to the broken lines. Pin and baste ¼ inch from the raw edges. Insert a cardboard square into the other pockets and baste the side edges.

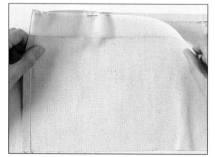

4 Fold the support along the broken line at the back of the shelves, enclosing the raw edges. Pin and stitch along the basting. Stitch the sides of the shelves in the same way, continuing the stitching to the front hemmed edges.

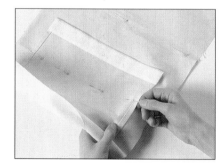

5 Cut one rectangle for the hanging strap 8 x 11½ inches. Stitch a ⅝-inch hem on the short edges. Stitch close to the first pressed edges to hem. Press under ⅜ inch on one long edge. Stitch a touch-and-close strip along one pressed edge. Turn the strap over and attach the other strip to the opposite long edge.

6 Cut a canvas roof and base 12⅝ x 25¼ inches. Mark the roof center widthwise with pins. Pin the strap centrally to one half with the long edges parallel to the pinned line. Stitch the strap ⅜ inch each side of the center. With right sides together, fold the roof and base widthwise in half. Stitch the short edges. Clip the corners and turn right side out.

7 Topstitch ¼ inch from the folded edge of the roof and base. Slip a cardboard square inside. Secure the roof layers together with three metal paper fasteners along the center of the strap. With right sides together, pin and stitch the roof and base centrally to the top and bottom of the support. Press under ⅜ inch on the raw edges of the support.

8 Fold the roof and base along the seam with the wrong sides together. Stitch ¼ inch from the seam, starting and finishing ¼ inch inside the finished edges. Pin the pressed edges and sides of the roof together. Stitch close to the outer side edges, then ¼ inch inside the edges, starting at the stitching on the back edge. Repeat for the base.

Belt and tie hanger

THIS VERSATILE hanger keeps belts and ties untangled and hanging in the closet, for easy selection when you're in a hurry. This storage idea relies on used wood that can be resized easily for new projects. Reclaiming seasoned, used woods has many advantages besides cost, since the wood will not warp or twist once it is recut.

1 Select a suitable piece of wood 1¼ inches thick. Using a carpenter's square against a straight-cut edge, mark, then cut the wood 2¾ x 12½ inches. In the picture, a reclaimed rabbeted door frame is being downsized along its length.

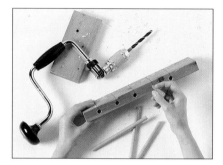

2 Following the template on p.134, drill five ⅜-inch dowel holes in the sides, using a self-starting auger bit in a brace. Carry on through the workpiece into a scrap of lumber to achieve a clean exit hole. Check that the fit is reasonably loose by inserting a length of dowel.

3 Cut five 10¾-inch lengths of ⅜-inch diameter dowel. Lightly sand the ends and paint them white. As the dowel rods are not glued into the body, they can be replaced with different lengths if you want to change the use of the hanger at a later date.

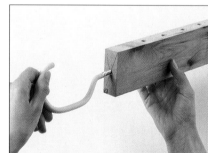

4 Select a large screw hook and draw corner-to-corner diagonals on the top end of the hanger. Drill a suitable pilot hole where the lines cross and screw in the hook. Sand and varnish the hanger, then hang it up to dry. Finally, insert the dowels.

Clothing envelopes

THESE PLASTIC-FRONTED envelopes make their contents visible and keep them clean and safe. Slip sweaters, T-shirts, and underclothes into the envelopes to protect them in your suitcase when you are traveling, or for storing delicate or out-of-season clothes in a drawer or closet.

1 Refer to the diagram on p.134 to cut a clothing envelope back from fabric and front from transparent plastic. Pin the front to the back close to the outer edges, leaving the upper edge open.

2 Turn under one end of 1-inch wide bias binding to start. Open out one edge of the binding and stitch along the fold line, making a ⅜-inch seam allowance on the envelope. Cut off the excess binding. Fold the binding in half, enclosing the raw edges. Baste in place.

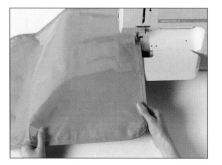

3 Topstitch close to the pressed edges of the binding.

4 Following the manufacturer's instructions, attach a gripper to the flap. Slip a sweater or a couple of T-shirts into the envelope to judge the position of the corresponding gripper and attach it to the plastic.

Blanket chest

THIS HEAVY-DUTY storage chest is a traditional favorite for storing blankets and linens, but it could also be used to store a variety of things around the home. This box is made from three different woods and has a mixed paint and varnish finish to suit the woods used.

The blockboard sides of the blanket chest have been finished in an oil-based paint for durability, and the mitered wood has been varnished. The lid has had a black stain run into the grain before sanding and varnishing. Attach the lid with two hinges and mount a catch on the front of the lid to lift it. Refer to the template on page 134 as a guide.

1 The sides of the box are cut from ¾-inch blockboard. Cut two panels for the front and back 12¾ x 30 inches and two side panels 12 x 12¾ inches. To join the corners, hold up a board to its matching piece and mark a half rabbet cut with a marking knife on the short edges of the front and back panels. Cut the half rabbets and slot in the side panels to check the fit. Glue inside the rabbet and pin the joint with 1-inch brads. Counterpunch the heads, then fill, sand, and prime the surfaces.

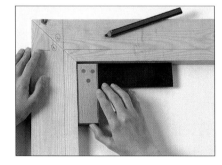

2 From 1 x 4-inch board, cut one 34-inch length, one length a little longer than 34 inches to allow for mitering and two 16½-inch lengths. Miter the ends and position the pieces to form a rectangle. Mark a track to support the box base 1 inch from the inner edges, then another track ¾ inch wide for the blockboard panels.

3 To form the mitered base, attach each mitered piece to the underside of the box by gluing and screwing through and countersinking 1½-inch countersunk screws. Take care to line up the mitered angles accurately. Do three sides, then offer up the final longer piece, cutting and adjusting the final miter if needed by marking with a knife, using a combination square.

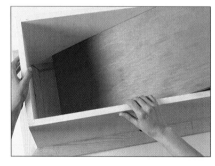

4 Cut the box base 11¼ x 28½ inches from ¾-inch blockboard. Run wood glue along the track marked for the box base. Lower the base into the box, resting it on the glued lip. Carefully place a weight, such as a bucket of water, inside.

5 Cut and miter 1 x 2-inch wood to make the mitered top lip for the box, using the same method employed to attach the base. Position the mitered pieces into a rectangle 14¾ x 31⅞ inches, the inner edge being flush with and an extension of the inside edge of the box panels. Glue and screw the lip to the box with 1½-inch countersunk screws, in the same way as you attached the mitered base.

6 From 1 x 1-inch wood, cut a 28½-inch length for the front support for the lid. Check the fit under the top lip, even with the top of the front panel. Glue in position and clamp with C-clamps. Wipe off any excess glue with a damp cloth and leave to dry.

7 Make the lid from blockboard and finish with an edging strip. The finished size of 11¼ x 28½ inches must have the width of the edging strip deducted from all four sides before the blockboard is cut. Tack and glue the edging strip onto the blockboard edges, using 1-inch molding tacks. Overlap the ends, cutting and sanding them to achieve an exact fit.

the bathroom

A small oasis

The bathroom should be a private sanctuary where you can unwind and pamper yourself. Most bathrooms are small, and although you may not need to store any large items, numerous toiletries including shampoo bottles, toothbrushes, razors, and make-up can quickly become cluttered, and soggy towels will need a place to dry.

Cabinets and shelving

Because of the size of the bathroom, wall space and corners need to be used for maximum storage. Open shelving need not be dull if you store your toiletries with a little imagination: pour perfumes and lotions into attractive glass bottles and display them to give a sophisticated atmosphere, or keep spare bars of soaps in a small wicker basket interspersed with seashells. Before installing a new cabinet or shelving, think carefully about the positioning. If sited over a sink, lean over the sink and raise your head slowly; make sure your head is not likely to come into contact with anything. A corner cabinet is indispensable for locking away medicines and first aid behind closed doors out of a child's reach, while remaining easily accessible to adults for safety and speed in an emergency.

Hang a ceiling rack on a pulley to store loofahs and brushes. Pot plants sited here will benefit from the room's steamy atmosphere and leave lower surfaces clear. Wall-mounted holders for soap and toothbrushes will also leave surfaces clear for toiletries in use. Alternatively, put a toothbrush holder and toiletry caddy inside a cabinet door. The equipment used for a wet shave needs to be well-ventilated and to have a slatted tray to drip into.

To make additional cupboard space, a vanity unit built under the sink will provide extra storage and a sliding panel at the side of the bathtub can hide cleaning materials. Make these an attractive feature of your bathroom in stripped pine or mosaic tiles. If your cupboard has a drawer, use a flatware tray (see pages 40–41) to hold small items like make-up.

right: **A tall, slim trolley shelving system with wire baskets and a towel rod creates a versatile and compact storage area to hold all bathroom accessories together in one place.**

below: **Use a traditional bath rack in wood or chrome for soap, washcloths or children's bath toys—or even to rest a book or magazine while soaking in the bathtub.**

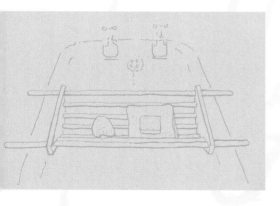

Showers and baths

Shower units are often badly designed with regard to storing the items you will need to have at hand when showering. Remove a wall tile in the shower and replace it with a tile that has an integral soap dish or install a plastic suction tray that can be easily removed and cleaned.

A soak in the bath is often a bit of an indulgence, but you still need to have your loofah within easy reach. A chrome or wooden rack across the bath can hold bathing accessories, and some even have a book-rest for the ultimate indulgence. If you are designing a bathroom from scratch, build in a "shelf" to the depth of a tile around the bathtub. This will provide a home for bottles and soaps so they won't have to balance precariously around the edges.

Wet, crumpled towels draped around the bathroom are an unattractive sight. Invest in a heated towel rod, so towels can be dried quickly and folded away. To create additional drying space for laundry as well as towels, hook a short rail over the top of the shower rod, so they can drip into the tray or bath.

Many parents like to make bathtime a fun time for children, to encourage cleanliness. To keep bath toys neat hang them in a plastic net bag on a suction hook in the shower or the side of the bathtub. Hang a plastic-lined drawstring bag on a hook for each member of the family's own toiletries. Children will be encouraged to clean up after themselves if they have their own personalized bag.

On the practical side, always bear in mind that steam is propelled boiling water, so uncovered screw heads will rust unless they are capped or sealed. When you are assembling and attaching shelves or units in the bathroom, always use waterproof glue.

right: **A purpose-built shelf uses the dead space above the lavatory to store clean towels and toiletries.**

Pegged shelf

THIS COUNTRY-STYLE shelf has been created from reclaimed wood, with two rows of pegs underneath to hold hand towels, bags, or a bathrobe, and the shelf making use of the space above to house toiletries or bathroom accessories. Shelves are always versatile throughout the home—this useful storage item could be used in a child's bedroom to hang a school bag or a jacket, or used as a small coat rack in a hall, with room for gloves or scarves on the shelf above.

Always remember to use waterproof glue when making wood joints that are exposed to water and steam.

1 Cut a wall piece 20 inches long and a shelf piece 21½ inches long from 1 x 4-inch wood. Cut an underside support piece 15 inches long from 1 x 2-inch wood. Cut a 1½-inch angled corner at both ends of the wall piece and a ¾-inch angled corner from both ends of the underside support. Round off the two outer corners on the shelf.

2 Mark the six drill holes on the wall piece and the two drill holes on the underside support as shown on the diagram on p.134 and drill six ⅜-inch dowel holes at the marks on the wall piece and two ½-inch dowel holes at the marks on the underside support. Insert a dowel into the holes to check the fit.

3 Cut ⅜-inch dowel into two 3-inch lengths, two 2½-inch lengths and two 2-inch lengths for the wall piece. Cut two 3-inch lengths of ½-inch dowel for the underside support. Smear wood glue around the dowels at one end and push into position. Wipe away excess glue at the front and back. Lightly sand and round the ends.

4 Screw the underside support to the wall piece, centering it even with the top edge, using 1¼-inch countersunk screws through the back.

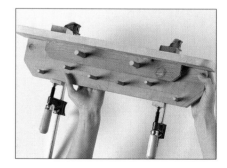

5 Center the shelf on top of the joined wall and underside support pieces and glue into position. Clamp both ends of the shelf with large C-clamps or bar clamps and wipe off any excess glue with a damp cloth. Set aside to dry. Sand the finished piece and apply four coats of varnish.

Laundry bag

WASH-DAY need not become a chore with this sturdy portable laundry bag. This tailor-made drawstring bag is designed to line a large wicker basket and can be lifted out to carry to the washing machine.

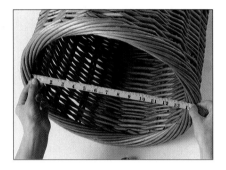

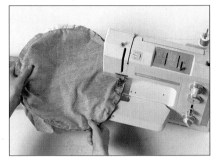

1 Measure the diameter of the inside of the wicker basket. Use a compass to draw a circle the same size on paper. Carefully measure the circumference. Add a ⅝-inch seam allowance. Cut out and use as a pattern to cut a circle for the base from fabric.

2 Cut a rectangle or square of fabric for the bag, two sides measuring the length of the circle circumference plus 1¼ inches and the opposite sides measuring the height of the basket plus 9½ inches. With right sides together and making a ⅝-inch seam allowance, stitch the height of the basket edges together. Press the seam open and finish the raw edges. Pin and stitch the base to the lower edge. Finish the seam.

3 Press ⅜ inch, then 1½ inches to the inside of the upper edge. Stitch close to both pressed edges.

4 Following the manufacturer's instructions, insert an even number of eyelets equidistant apart and ⅜ inch below the upper edge. Thread with cord, knot the ends, and unravel the cord below the knots.

Tiered shelf unit

MAKE A practical unit to support your own metal, plastic, or wicker trays that will stand on a shelf or bathroom windowsill and hold small toiletries and other bathroom accessories. This unit is designed to take two trays that are 2¼ inches deep and one that is 4 inches deep. Remember to adjust the measurements given to suit your own trays if they are different to the ones used here.

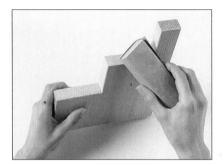

1 Cut two side pieces 7¼ x 8½ inches from 1 x-8 inch wood. Cut out two pieces 2½ x 3 inches and 3 x 5 inches from the two sides, referring to the diagram on p.135. Sand the edges.

2 Cut two base pieces 8½ inches long from ½ x 2-inch wood. Round off the corners with sandpaper. Drill two holes through the center base line and screw to the bottom of the side piece, centering it using 1¼-inch countersunk screws.

3 Cut three upper cross pieces 1¼ inches wide and the length of your trays, and one lower cross piece the length of your trays minus the width of the two side pieces, from 1 x 6-inch wood. Half rabbet both ends of the upper cross pieces to overlap the side pieces at the front of the steps. Glue and pin in position with 1-inch brads.

4 Measure between the two base pieces that now act as "feet" for the sides, and cut to size a connecting base piece from ½ x 2-inch wood, to span the gap. When this piece is positioned, even with the back of the uprights, place the lower cross piece on top. Glue the pieces together, then secure to the sides by driving 1¼-inch countersunk screws into the lower cross piece ends from the outside of the upright.

5 Sand and fill any defects in the wood and wood joints to achieve a smooth finish. Clean with mineral spirits. Spray paint the unit. Screw in hooks to hold the trays in place.

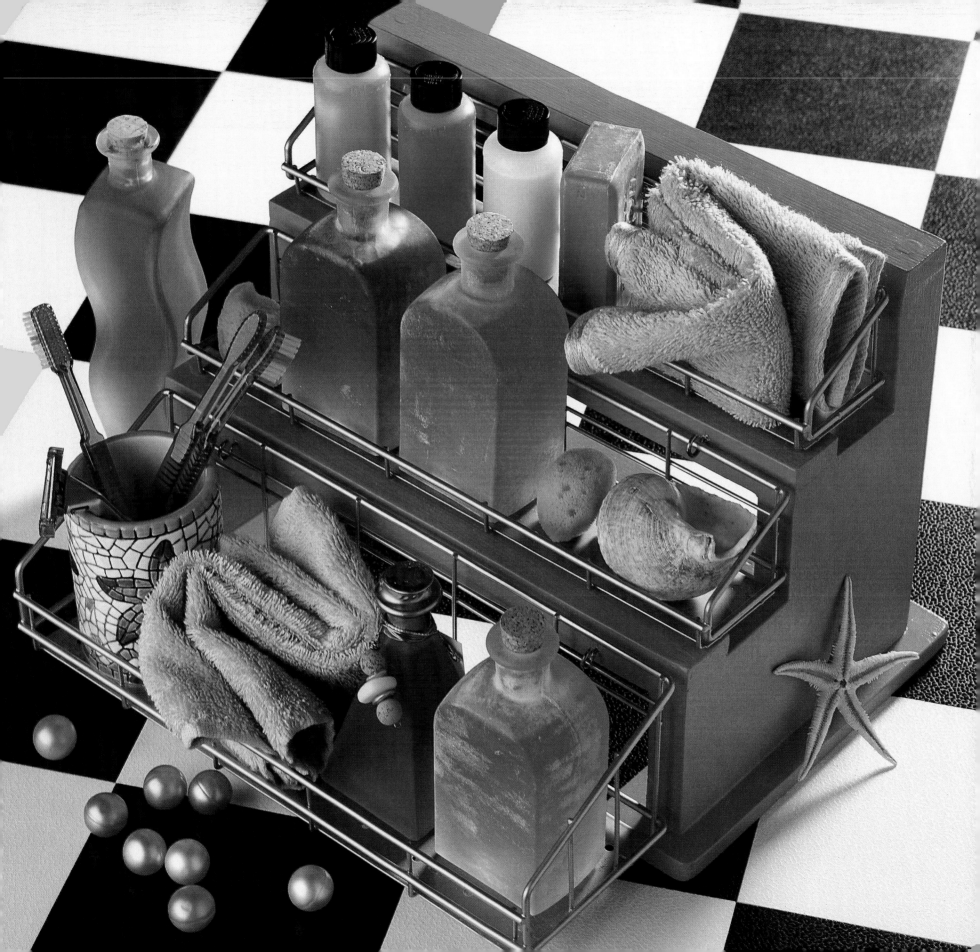

Toiletry pockets

IF YOU have a very small bathroom without a spare chest or shelf space to store toiletries, this handy organizer is ideal, as it can be hung from the back of the door. The robust pockets are made of plastic-coated fabric, which means that any spills can be easily wiped away.

This is a very versatile storage idea that could also be used to hang on the back of the bedroom door to keep socks or undergarments neat, or in the children's room to store small toys or stuffed animals.

1 Refer to the diagrams on p.135 to cut one lower pocket strip, one middle pocket strip, and one upper pocket strip from plastic-coated fabric. Press under ⅝ inch on the upper edges and stitch in place to hem the pockets. To make pleats, fold the strips along the solid lines to meet the broken lines and press. Baste in position along the lower edge.

2 Press under ⅝ inch on the short edges. Cut two rectangles of fabric 21 x 30¾ inches for the front and back. With right sides together, pin the lower edge of the lower pocket strip centrally 2 inches above the lower edge of the front. Stitch, making a ⅝-inch seam allowance on the pockets.

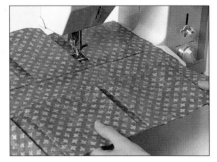

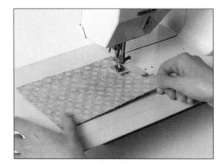

3 Lift the pocket strip and pin the short, pressed ends parallel with the long edges of the front. Topstitch close to the ends and lower edge of the pockets. Topstitch again ¼ inch inside the first stitching. With right sides together, pin the lower edge of the middle pockets centrally 11⅝ inches above the lower edge of the front.

4 With right sides together, pin the upper pockets centrally 21 inches above the lower edge. Stitch the middle and upper pockets, making a ⅝-inch seam allowance. Lift the pockets and topstitch as before. To form separate pockets, stitch between the broken lines of the pleats. Stitch back and forth a few times at the top of the stitching to reinforce.

5 Cut six rectangles of fabric for the flaps 4¾ x 7 inches. With right sides together, stitch the flaps in pairs, making a ⅝-inch seam allowance and leaving one long edge open. Clip the corners and turn right side out. Press, then topstitch close to the stitched edges. Topstitch again ¼ inch inside the first stitching.

6 With right sides together, pin the flaps edge to edge to the front with the raw edges 1¼ inches above the middle pockets. Stitch, making a ⅝-inch seam allowance on the flaps. Flop the flaps over the middle pockets. Topstitch ¼ inch from the upper flap edges.

7 With right sides together, stitch
the front and back together
making a ⅝-inch seam allowance and
leaving a 17½-inch gap in the middle
of the upper edge. Clip the corners.
Turn right side out and press, pressing
the open edges to the inside.
Topstitch close to the sides and lower
edges, stitch then ¼ inch inside the
first stitching. Stitch across the fabric
2⅜ inches below the top.

8 Slip a 17½-inch strip of ³⁄₁₆ x 1-inch
wood into the opening to rest on
the last row of stitching. Topstitch
across the upper edge close to the
pressed edges, stitch then ¼ inch
inside the first topstitching. Attach an
eyelet ¾ inch inside the outer edges at
the corners on the upper edge to
hang on the hooks.

Cleaning box

UNSIGHTLY HOUSEHOLD cleaning materials, sponges, and cloths can be stored neatly out of sight in this useful box. The sliding cushioned lid means the storage box can also double as a handy bathroom seat, disguising its practical purpose. As an alternative, it will also make an excellent place to put away toys in the living room, with the seat covered in fabric to match the decor.

To create a smooth finish, use 1¼-inch countersunk screws, countersinking and filling all the heads. Refer to the diagram on page 136 as a guide.

1 Cut two side panels 9½ x 18 inches from ¾-inch blockboard and edge the shorter edges with ¾-inch edging strip. Cut two end panels 9½ x 10 inches from ¾-inch blockboard. Cut four base and lid supports 18¼ inches long from 1 x 2-inch wood. Screw the supports to each long edge of the side panels, flush with the outer surface of the blockboard.

2 Cut four pieces 9½ inches long from 1 x 1-inch wood. Measure in 1 inch from the shorter side panel edge, not including the edging strip, and screw to the panels parallel to the edge.

3 Screwing through the adjoining edge of the 1 x 1, attach the end panels, forming a box shape.

4 As the base supports overlap the blockboard on the inside (see Step 1), they form a lip that holds the base in position. Cut the base from ¾-inch board 10 x 12¾ inches and glue in position resting on the supports.

5 Cut the sliding lid from ¾-inch board 13½ x 20½ inches. Cut two supports 18½ inches long from 1 x 2-inch wood, positioning them on the underside of the lid so they will slide between the lid supports. Cut a corner on the inside of the wood strips. Screw through the board into the strips to anchor in place.

6 Cut a piece of 1½-inch thick foam to fit the lid and glue in place. Rest the lid on the fabric. Working out from the center of each long side, staple or use tacks to secure the fabric to the underside of the lid. Fold the corners under neatly and attach the short edges of the fabric to the lid the same way.

Slatted towel rack

THIS SIMPLE slatted device is the modern way to hold towels. It is designed to take neatly folded or rolled spare towels or washcloths, ready for use. The open design allows air to circulate, and the rack can either be used freestanding, or it can be wall-mounted, as shown here, in order to take up a minimum of space in the bathroom.

1 Cut a wooden upright 1¼ x 4 x 17 inches. Shape the top toward a point and sand to a smooth finish. Refer to the diagram on p.136 to mark the wood for cutting.

2 Cut the sides of a ⅜-inch rabbet channel in four places, following the diagram, to a depth of ½ inch. Chisel away the waste.

3 Cut eight slat support pieces 6 inches long from 1 x 10-inch wood. Each piece has a 45 degree angle front end cut facing upward. Cut a half rabbet at the left side of four pieces and at the right side of the remaining four pieces. The half rabbet is the depth of the upright: 1¼ inches.

4 Cut eight pieces 9 inches long from ⅜ x 2-inch wood for the slats. Sand the edges and position in the cut rabbet on the upright to check the fit.

5 Attach the slat supports to the sides of the upright even with the rabbet's lower edge using 1¼-inch countersunk screws in brass seatings. Position a slat in each rabbet, resting on the supports, leave a 1-inch gap, then position the remaining slats on the front of the supports. Glue and tack into position with ¾-inch molding tacks.

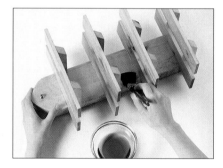

6 Set aside to dry overnight. Sand the complete unit with fine sandpaper. Remove any marks with mineral spirits and apply four coats of clear varnish.

the children's room

Fun and games

Tidiness is not a natural characteristic of most children, who often view their bedroom as a lawless zone, free from the restriction of grown-ups' rules! To encourage them from an early age to put away their toys and clothes, storage must be fun and easily accessible. Kid's storage also needs to grow with them and have the flexibility to adapt to their later lifestyles, as well as being robust enough to take heavy wear and tear.

Involve the children in the design and choice of colors when planning the room, then they will take pride in the result. The drawstring bags and pen and pencil holders on pages 108–111 are simple enough for children to help make.

Encouraging neatness

Choose a chest of drawers with shallow drawers for clothes or as a desk. These are easier to keep tidy than having to burrow to the bottom of a deep drawer in search of something, turning everything upside-down in the process. Paint a picture of the contents on the outside of the drawer, so that everything can be put in its place at the end of the day. Paint the sides of a chest with blackboard paint for youngsters to draw on—it may preserve your walls and can be repainted when the child is older. A low-cost storage system can be created with a basic set of freestanding shelves, using transparent plastic crates as drawers. The crates are lightweight to lift out and move around, and the contents are visible to avoid endless rummaging for missing belongings.

Encourage clothes to be hung up in a closet by buying character hangers, or making a game out of clearing up by having different-colored plastic hangers for different items of clothing. Hooks attached to the back of the bedroom and closet doors can be used to hang coats, hats, and bags. Take care that they are not low enough so they are bumped into or so high that the child cannot reach.

Creating a play space

Children love the idea of a platform bed—as well as creating a lot of space underneath, it gives masses of inspiration for play too. Before installing a new bed, take time to find out what kind of play area your child would like—perhaps a desk area with spotlights to play games on the computer and do their homework, or maybe a secret den with pillows inside and curtains to pull across, or a couch to sit on with friends and watch cartoons on TV.

Having to share a room with a sibling can often cause friction. A screen or bookcase can be used for storage and as a room divider for children sharing the same space; it will give some privacy and a sense of territory. Choose adjustable shelving that can be altered as the children mature. Take care that young children cannot use the shelving as a ladder. If there is a large age gap between children sharing a room, install low

below: **A platform bed designed for an older child has an extensive desk area underneath the bed for doing homework or playing computer games.**

some shelves that the youngest can have access to and higher shelves out of their reach for the older sibling's treasures.

Be imaginative when it comes to kids' storage. Suspend colorful baskets or painted buckets on plastic chains to hang below overhead cupboards or shelves to store small toys. Clip stuffed toys garland-fashion across a wall on colored ribbons or nestle them in a hammock hung across a corner of the room.

A pinboard is fun for children of all ages and can hold favorite pictures and their own artwork while discouraging them from pinning and sticking things directly onto the walls.

Certain games can last for days and take up a lot of floor space. Put a castor on each corner of a wooden board to hold a small train set or a jigsaw puzzle, then wheel it out of sight under the bed until it is ready to be played with again.

Large toys such as a tricycle or go-kart take up a lot of space and often cannot be stored outdoors. Support them on a wall on two shelf brackets under the front axle. Extend the brackets with wooden strips if necessary and cover the support with foam so they do not damage the paintwork. Bicycles and tricycles can be hoisted to the ceiling to make more floor room.

Fabric screen

CHILDREN WILL love this multipurpose screen—perfect for hiding a messy corner of the room when mom or dad want you to clear up or for dividing a bedroom shared with a brother or sister to create a private space. Strong wooden frames support the three colorful panels with their roomy pockets. The large purse can be taken off for play, and the practical panels are easy to remove for laundering.

Use a mediumweight fabric that will be strong enough to hold the contents of the pockets and the purse. Make a ⅜-inch seam allowance throughout.

1 Cut six pieces of 1 x 2-inch wood 48 inches long, six pieces 18 inches long and six pieces 16½ inches long. Cut each corner as a half overlap joint. Rabbet one half and use as a guide for the other to ensure a tight fit.

2 Position the 16½-inch support piece to help form a 90 degree corner ½ inch from the frame edges and glue and clamp the joint using C-clamps. Repeat for all the corners, assembling the last two corners of each frame at the same time. Set aside overnight, then sand and apply four coats of clear varnish. Join the frames with flush hinges.

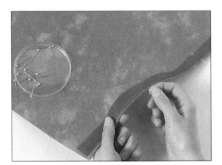

3 Cut six rectangles 18 x 48 inches from fabric for the panels. With right sides together, stitch the panels in pairs, leaving a gap on the lower edge to turn. Clip the corners, turn right side out and press with the opening edges inside. Stitch touch-and-close tape close to the short edges. Attach the corresponding strip to the screen support pieces with tacks.

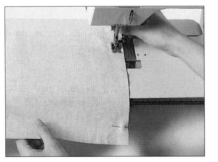

4 Refer to the diagrams on p.136 to cut two pocket strips and a purse front from fabric. Stitch the pocket strips together on the long upper and short side edges with right sides facing. Clip the corners and turn right side out. Press and baste the raw edges together. Stitch a 1¼-inch deep hem on the purse front.

5 To make pleats on the pocket strip and purse front, fold along the solid lines to meet the broken lines and press. Baste across the lower edges. Cut two rectangles of fabric 9⅞ x 15 inches for the purse back and lining. Pin the front to the lining with right sides uppermost and matching the raw edges.

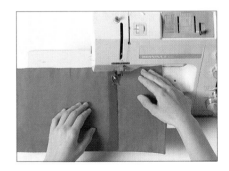

6 Pin the back on top with right sides facing. Stitch the outer edges, leaving an opening to turn. Clip the corners. Turn right side out and press with the open edges inside. Stitch the soft section of touch-and-close tape 4¼ inches below the upper edge of the back. Work two rows of topstitching on the outer purse edges. Fasten the purse with a snap.

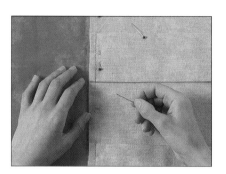

7 With right sides together, pin the lower edge of the pocket strip centrally across the middle of one panel. Stitch. Lift the pockets and pin the short edges parallel with the sides of the panel. Stitch along the center between the broken lines to separate the pockets.

8 For patch pockets, cut two pieces 7 x 10 inches and two pieces 7 x 8 inches. Make a 1½-inch deep hem on the short upper edges. Press under ⅝ inch on the raw edges. Arrange the purse and pockets on the panels. Stitch the matching purse touch-and-close strip to the panel. Work two rows of topstitching on the sides and lower edge of the pocket strip and patch pockets.

Toy cupboard

OCEAN WAVES lap around the shelves of this practical toy cupboard. Store games on the top shelves and stack those awkward, bulkier toys below, hidden from view when they are not in use by a blue roll-up shade. To create a different theme, choose another style of edging and a different color roll-up to match.

Make a couple of drawstring bags for additional storage to hang from the hooks that have been inserted under the middle shelf (see the instructions on page 108).

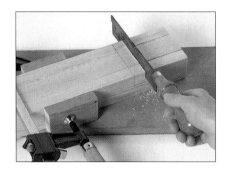

1 Cut four uprights 48 inches long from 1 x 2-inch wood. To cut a set of half rabbets in the top corner, another set 7 inches down, then a set 3 inches from the lower ends, clamp the uprights with bar clamps, mark the cut with a marking knife and carpenter's square. Use a small tenon saw to cut four rabbets simultaneously.

2 Cut three rectangles of blockboard 12 x 16 inches. Cut a matching rabbet for the uprights in the corners of each board, using the half rabbets already cut in the uprights as a template.

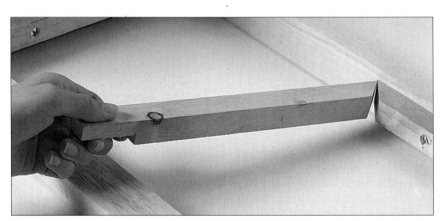

3 Start to assemble the basic frame by matching the halving joints together. After checking each individual fit, glue and screw the joint through the side, countersinking 1¼-inch countersunk screws. Start with the top shelf, attach the lower shelf next, and finally the middle shelf. Note that the uprights are side-on to the rear and front-on to the front, as shown in the diagram on p.137.

4 Cut four strips 11¼ inches long from 1 x 1-inch wood for the side supports, half-rabbeting one end of each strip to 1¾ inches to accommodate the side-on rear uprights. Screw into position under the middle shelf and on top of the base shelf ½ inch inside the shelf edge, countersinking 1¼-inch countersunk screws.

5 From the same 1 x 1-inch wood, cut two strips 26 inches long and miter one end of each for side supports. Position the strips to form a right angle with the lower shelf strips and screw into place behind the front uprights ½ inch from the outer edge, again countersinking 1¼-inch countersunk screws. Miter two more strips from the same wood 11¼ inches long and join at right angles to the 26-inch strips at the miter. Rabbet the other ends of these cross pieces to 1¾ inches and screw to the face of the rear uprights, countersinking 1-inch countersunk screws.

6 Cut two side panels from ½-inch plywood 9½ x 36⅝ inches. Make sure the corners are square. Mark the position of the support cross piece on the inside of the side panels and drill pilot holes. Using 1-inch screws with brass seats, secure the side panels to the support strips.

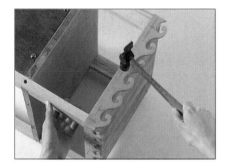

7 Cut and tack an edging strip to the front and sides of the top and middle shelf (a 2-inch deep wave-shaped strip was used for this project). Varnish the side panels and shelves, and prime and paint the rest of the wood. Cut a roll-up shade to size and mount it on the front of the cupboard below the middle shelf. Screw cup hooks under the middle shelf for extra hanging space.

Tented closet

THIS UNUSUAL and colorful piece of furniture has its origins in beach resorts of yesteryear. The fabric cover is supported by a sturdy wooden frame and allows access to a hidden shelf above the clothes. Roll up the "door" and fasten it with large buttons to reveal the closet's contents.

Follow Steps 1–6 on pages 102–103 to construct the basic frame; the side panels are optional here. Position a clothes rod centrally under the middle shelf, parallel to the sides. The rod and relevant hardware are available at hardware stores and can be cut to size with a small hacksaw. Attach one anchor, screw in position, slide in the rod and second anchor, and secure with screws. Make a ⅝-inch seam allowance throughout.

1 Cut two rectangles of plain fabric 4¾ x 18 inches for the straps. With right sides together, fold lengthwise in half and stitch the long edges and across one end. Clip the corners and turn right side out. Press the straps and work a buttonhole ⅝ inch from the finished ends.

2 Cut one rectangle of solid fabric and one of print fabric 16⅛ x 46⅞ inches for the door. With right sides together, stitch the rectangles on the long and lower short edge. Clip the corners. Turn right side out and press. Pin each strap to the upper raw edge 2¾ inches in from the long edges on the print-fabric side of the door.

3 Cut two rectangles of print fabric for the pediments 3½ x 17¾ inches. With the print fabrics uppermost, pin the upper edge of the door centrally to one long edge of one pediment. Cut two rectangles of print fabric for the front borders 6 x 47¼ inches. Press in half lengthwise with wrong sides together.

4 Matching the raw edges, baste the short upper edges of the borders to the pediment on top of the print-fabric side of the door. Pin, then stitch the other pediment on top. Turn the pediments right side out.

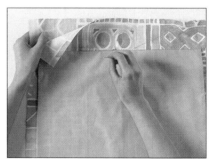

5 Press, then topstitch ¼ inch from the seam. Baste the raw edges of the pediments together. Cut print fabric for the cover 42½ x 49⅝ inches. With right sides together, stitch the long edges of the cover and front borders starting ⅝ inch below the upper edge. Finish the seams and press open.

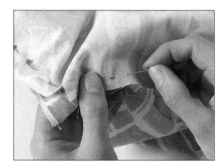

6 Cut a rectangle 13⅝ x 17¾ inches for the roof. With right sides together, pin the upper edge of the pediment centrally to one long edge of the roof. Pin the upper edge of the cover to the other roof edges, clipping the cover so it lies smoothly at the corners. Stitch, pivoting at the corners.

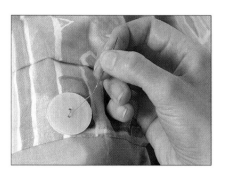

7 On the right side, sew a button to the pediment above the top of each strap. Slip the cover over the wooden frame. Pin the hem in place. On the front border seam, mark the position of the top of the middle shelf and underside of the lower shelf with a pin. Remove the cover. Stitch the hem in place. Cut four 20-inch lengths of cotton tape and mark the centers with a pin.

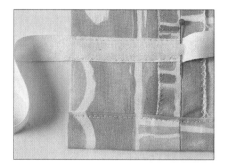

8 Match the tape pin to the seams above the position of the middle shelf and below the lower shelf. Pin one half of the tape across the front borders and stitch close to both edges of the tape. Press the front borders toward the inside along the long seams. Replace the cover and tie the tapes around the wooden uprights.

Pull-along toy box

A WAGON-STYLE toy box is sure to be a great favorite with all children. The brightly colored seat slides open to reveal hidden treasures inside. There are protruding metal eyes on the front of the box that help to guide it when it is pulled along by the attached rope. If you feel these are unsafe for your young children, replace them with short eyes or omit the eyes and rope altogether—the trolley can be pushed along by its rider's feet. Refer to Steps 1–4 on pages 92–93 to make the basic box.

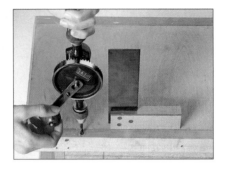

1 Sand and fill the box. To attach the wheels, select a bolt that will pass through the wheel hub hole and is long enough to pass through the base support and be secured on the other side. Drill a pilot hole for the bolt that will enable the thread to grip the sides as if it was a screw. This will stop the bolt from moving in the hole and allow the hub to rotate freely. Use a carpenter's square to ensure your drill is at right angles.

2 Screw the bolt through the wheel, then the base support. Make sure any hub lip is on the inside. When the bolt is in position and the wheel rotates freely, attach in the following order: a conventional washer, a spring or locking washer, and the nut.

3 Repeat this procedure until all the wheels are in position. Threading the bolts through the support is accomplished more easily by smearing on a small amount of grease.

4 Screw two large eyes into position on the inside front uprights at the front end of the toy box 2 inches from the base. Pass a rope through both eyes and knot it. Check that the box runs smoothly. Make and cover the lid, referring to Steps 5–6 on p.93. Paint the box, and thread some large, colorful beads onto the rope.

Drawstring bags

THESE SIMPLE bags with their large see-through pockets are a contemporary version of traditional drawstring bags. The transparent pockets help children to find small items that can easily go astray.

Not only are these bags cheap and very easy to make, but you will find that they rapidly become useful throughout the home, for laundry, storing spare dishcloths and plastic bags in a cupboard or in the kitchen, or to hold shoes in a suitcase to protect your clothes when you travel.

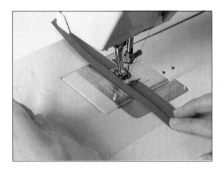

1 To make a pocket, cut a rectangle of transparent plastic 4¾ x 5½ inches for a small bag and 5½ x 6¾ inches for a large bag. Open out one edge of ⅝-inch bias binding. With ⅜ inch of the binding extending at each end, stitch along the fold line of the binding to one long edge of the pocket, making a ¼-inch seam allowance.

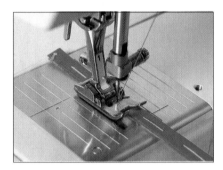

2 Turn under the ends of the binding. Fold the binding in half, enclosing the plastic. Baste then topstitch close to the pressed edges of the binding. To make the front and back, cut two rectangles of fabric for the small bag 10½ x 15 inches, and for the large two-color bag cut two pieces of two contrasting fabrics 6⅝ x 16½ inches.

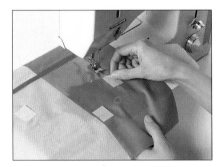

3 With right sides together and making a ⅜-inch seam allowance, stitch the contrasting rectangles in pairs along one long edge to make the front and back. Press the seam open. Tape the pocket to the front with masking tape at least 5¼ inches below the upper short edge. Topstitch close to the edges of the plastic, then ¼ inch inside the first stitching.

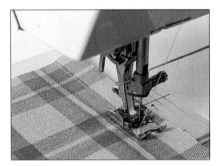

4 With right sides together and making a ⅜-inch seam allowance, stitch the front and back, leaving the upper short edge open and a ¾-inch gap 1½ inches below the upper edge. Clip the corners and press the seams open.

5 Press ⅜ inch, then 1 inch to the inside on the upper edge to make a channel. Stitch close to the first pressed edge.

6 Cut two 1-yard lengths of ribbon, attach one length to a safety pin, and thread through the channel, entering and emerging from the same hole. Thread the other ribbon through the hole on the opposite side of the bag. Knot the ends together. Adjust the ribbons to hide the knots in the channel.

Pen and pencil holders

A SET OF VIBRANT fun-colored holders will brighten up a child's desk and are perfect for storing pens and crayons. The containers have been simply constructed from sturdy recycled postage tubes and covered in a stylish fine foam, which has a soft, tactile finish. You will find this material available from many craft and hobby stores.

1 Use a craft knife to cut a slice from a cardboard postage tube for the holder. Draw around the circumference of the tube on thick cardboard and cut out the base. Glue the base under the end of the tube with all-purpose adhesive.

2 Cut a strip of thin, colored foam long enough to wrap around the tube and meet edge to edge and 1¾ inches taller than the height of the tube. Glue the foam around the tube with 1 inch extending above the upper edge. Glue the upper edge inside the tube. Trim the lower edge to ¼ inch from the base. Glue the trimmed edge to the base and decorate the holder with scraps of foam.

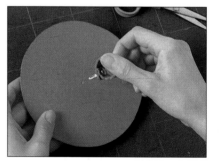

3 To make a lid, cut a circle of thick cardboard ¾ inch wider than the diameter of the holder. Cut a circle of foam 1½ inches wider than the diameter of the lid and glue together. Use pinking shears or decorative edging scissors to cut a strip of foam ¾ x 3¼ inches. Bend into a ring, push a brass paper fastener through the ends, then attach to the lid.

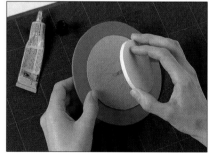

4 Cut a circle of foam board or corrugated cardboard 1¼ inches narrower than the diameter of the cardboard lid. Glue centrally under the lid. Trim the foam to ¼ inch from the cardboard and glue the trimmed edges under the lid.

the home office

Down to business

The home office is becoming an increasing reality for many people, as electronic communications allow business to carry on, whatever the location. Even if you don't work at home, you may need desk space for a computer or somewhere to file bills and important documents. A study or office in the home is often a luxury, and the home office sometimes vies for space with a spare bedroom, dining room, or an area for crafts and leisure activities, but commonsense organization and versatile storage systems can make it work successfully.

Dividing work and home

If you work from home full time, give high priority to assessing the needs of the home office. Then your work will be less likely to encroach on other areas of the home. Whatever size space you have to make use of, always be aware of the fact that over the years you are likely to need more storage and not less, and that projects which are no longer current will need to be archived, perhaps in an attic or cellar. Exercise good admin skills by making a point of sorting through paperwork regularly, before it becomes messy heaps cluttering your work surface. Label files and storage boxes accurately; it will save time and frustration when searching for that important document.

If the home office shares space with another room, a divider may be a sensible option; a partition with shelving on one or both sides would work well. Be careful not to allow work to invade your home life. A foldaway desk helps conceal a working environment, and a blind can be fitted over shelving holding files. A trolley with a fax machine or photocopier can be wheeled out of sight at the end of a working day.

Disguising the home office

If your office has to double up on occasion as a guest room, take time to consider attractive storage options, while at the same time giving priority to your own work space. For a start, invest in a sofa-bed to leave a maximum area free for your workstation. Convert the inside of a closet into a shelving unit to file books and binders, so when guests arrive the doors can be closed and locked, and the room becomes a bedroom. If you can't hide all the evidence of your working day, choose attractive storage boxes and colorful files, or use containers such as pottery or unusual baskets that will enhance the room. Spray paint your metal filing cabinet a lively color to beat those corporate fawns and grays.

To create a self-contained minioffice, use a large two-door cabinet with a deep shelf to use as a desk and put shelving above and a trolley of wire baskets below. A pull-out shelf under the desk will give additional space for holding reference material while you are working. As a guide, 28 inches is a standard desk height and a shelf 20 inches above the desk is the average maximum reach from a seated position. At the end of the day you can close the office and all your work materials will be instantly put out of sight.

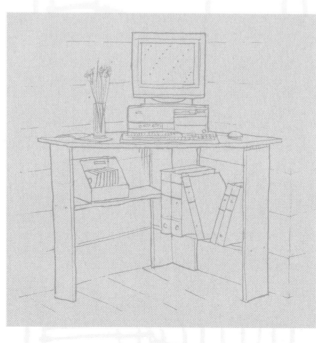

right: **The smallest area in a room can be used to create a work space. This corner desk is large enough to take a computer, with shelving underneath for some additional files.**

far right: **This versatile desk area can be used as a home office or for hobby activities. Storage is provided by a portable wire trolley and a stack of labeled boxes.**

Bulletin board

HAVING A PLACE to display notes of important events, schedules, or telephone numbers is a vital accessory in any work space. This traditional cork-style board has a contemporary twist, since it also doubles as a desk set. Practical detachable canvas pouches pinned to the lower edge will hold small office supplies that are otherwise easy to lose. This leaves the desk area clear and ready for work.

A bulletin board is a useful addition to many rooms. Use it in the kitchen as a reminder to pay bills, or in a child's room so they can pin up their own artwork without damaging the walls, keeping crayons in the pouches.

1 Glue four 12-inch square cork tiles to the rough side of a 24-inch square of masonite. Weight with heavy books and set aside to dry.

2 Cut a rectangle of canvas 3⅞ x 5⅜ inches for the pouch back and a rectangle 5⅜ x 9⅜ inches for the pouch front. Press ¼ inch then ¾ inch to the right side on one short upper edge of the back and one long upper edge of the front. Stitch close to the first lower pressed edge on the back. Finish the seams.

3 Open out the pressed edge on the front. With wrong sides together and the lower edges even, stitch the long back edges to the short front edges, making a ⅜-inch seam allowance. Press the seams toward the front.

4 Use the template on p.137 to cut a base from canvas. With right sides together, pin and stitch the base to the pouch, matching the dots to the seams and making a ⅜-inch seam allowance. Finish the seam.

5 Refold the front along the pressed edges. Stitch close to the lower pressed edges to make a channel. Turn right side out. Thread an 8¼-inch length of ½-inch wide plastic boning through the channel. Stitch across the ends of the channel. Pin the pouch to the bulletin board with thumbtacks.

Shelf compartments

CUSTOM-BUILD your own unit to fit the available space in your home office and give easy access to your work materials. The divided compartments help you separate and organize an assortment of different items. You can construct any size unit with varying compartments designed around the items you need to store, by using simple halving joints cut into ½-inch plywood as described on page 28.

1 Start by drawing a plan of compartment sizes for the items you wish to store within the external dimensions. Using 4-inch wide strips of ½-inch plywood for the depth of the compartments, cut the wood into lengths to match the width and height of the finished unit.

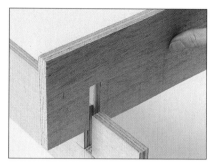

2 To form an internal compartment, cut and chisel ½-inch rabbets half the width of the pieces to be joined, following your plan. Slot all the pieces together, checking the fit as you go, to start to form the unit.

3 Form an external corner by cutting simple one-sided halving joints in the same manner. Check the fit again.

4 When all cuts and joints are completed, glue and clamp the design together. Cut a piece of ¼-inch plywood for the back. Glue the edges of the compartments to the back and secure with 1-inch molding tacks. Leave to dry, sand, and fill any rough edges or surfaces and paint with the finish of your choice.

Portable filing chest

UNFINISHED WOODEN furniture is inexpensive and widely available, so it is ideal to customize for your own room. This stylish, mobile set of drawers has been created from two purchased small chests joined together. Trolley wheels underneath allow you to move it to and from your workstation, or hide it away under the desk when not in use.

1 Remove the drawers from the chests and cover them with light oak-colored varnish. Leave to dry, sand lightly, and then varnish again.

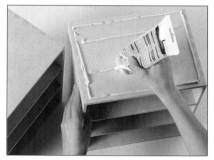

2 Glue the chests one on top of the other with wood glue or filling adhesive. If the base of the chest is thinner than ½ inch, miter and glue ⅜-inch thick strips of wood to the base at the front and sides. Weight the chest with heavy books overnight while the glue dries.

3 Varnish the chest with two coats of teak-colored varnish, sanding between coats. To position a castor or wheel with a top mounting, turn the chest on its back. Measure ½ inch in at both sides of each corner and position the castor within the marks. Mark all hole positions with a pencil.

4 You only need to attach the castor base by two corners, on a diagonal. Use a sharp awl to mark the starting point for the screws, which should be supplied in the pack with the castors. Screw in place. Remember if you attached mitered strips of wood to the chest, the corner screws at the front must be on the opposite diagonal to the miter.

Corrugated-plastic boxes

CORRUGATED PLASTIC is an excellent material from which to make these large storage boxes because it is durable and lightweight, making it ideal to store paperwork or hobby accessories. This colorful plastic "riveted" with paper fasteners gives these boxes a contemporary style, but you could go for a more traditional look by using corrugated cardboard.

1 Use the diagram on p.138 to draw the box and lid on corrugated plastic. Cut out along the solid lines with a craft knife. To make the box bend, score ⅛ inch on each side of the broken lines, breaking only the top surface of the corrugation. Mark the dots ½ inch inside the outer edges on the right side of the box and the lid.

2 Use the craft knife to recut the scored lines, holding the knife at a diagonal angle with the blade toward the broken line. Gently pull out the narrow strip of plastic, leaving a V-shaped channel.

3 Fold the sides up, tucking the tabs inside. Use an awl or scissor points to make a hole at the dots through the short sides and tabs to secure with paper fasteners.

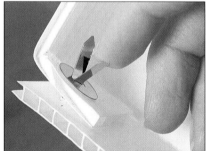

4 Insert the fasteners through the holes. Thread on a washer, if supplied, and splay the prongs open inside the box and lid. Trim the prongs if they are too obtrusive.

Portfolio

STORE ARTWORK and large documents in this traditional portfolio, which has been made from poster board covered in hard-wearing canvas and fastened together with ribbon. The strong cover and inner flaps keep all materials inside flat and secure.

Adapt the measurements given here to make a smaller portable folder for stationery or a larger portfolio for your favorite masterpieces. This portfolio is designed to be large enough to hold up to 11 x 16 inches.

1 Cut two rectangles of poster board 13½ x 18 inches and two rectangles of colored paper 14¼ x 19½ inches. Use spray adhesive to stick the papers to the boards, leaving ¾ inch extending beyond the short edges and one long edge. Glue the corners, then the edges to the wrong side with craft glue.

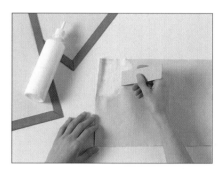

2 Cut a rectangle of canvas 8½ x 20½ inches. Apply glue to one side of the canvas, spreading the glue thinly and evenly with a glue spreader or strip of cardboard. Set aside to dry. This will stop the fabric fraying. Glue the end of an 18-inch length of ribbon to the middle of the covered long edge of both boards on the wrong side. Trim the ends of the ribbon diagonally.

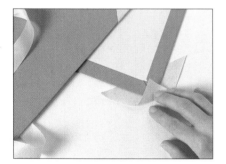

3 Refer to the template on p.137 to cut four triangles from the stiffened canvas. To reinforce the portfolio corners, use glue to stick a triangle to the right side of each covered corner with ¾ inch extending beyond the edges of the boards. Glue the corners, then the edges to the wrong side.

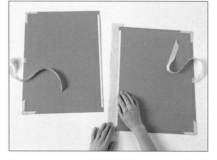

4 Cut two pieces of paper 12½ x 17½ inches to line the portfolio. Glue centrally to the wrong side of the boards with spray adhesive. Cut two strips of stiffened canvas 2 x 19¼ inches. To make a hinge, glue one strip along the uncovered edge, overlapping the right side of the board by ⅝ inch and with ⅝ inch extending beyond the ends.

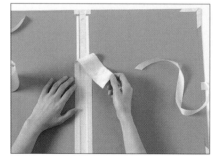

5 Glue the other edge of the hinge to the other board the same way. Glue the ends of the hinge to the inside. Glue the remaining strip along the hinge on the inside, sandwiching the boards. Cut off the excess canvas.

6 Cut four rectangles of colored paper 5½ x 12⅝ inches for each flap. Fold under ⅝ inch on one long edge of each flap for a tab. Stick the tabs to the boards close to the upper and lower edges on the inside of the portfolio with double-sided tape or craft glue.

Accordion file

A FOLDING file is an excellent way of putting your paperwork in order, including bank statements, bills, correspondence, recipes, and much more. The divided sections keep all your filed items separate and easy to find.

Although costly to buy, an accordion file is inexpensive and easy to make and can be covered with plain paper, as here, or with fabric. Choose colors to coordinate with your office for a streamlined look.

1 Refer to the diagram on p.139 to cut a strip of thin cardboard for the accordion. Score along the long lines ⅝ inch apart with the back of a scissor blade so you dent the board instead of cutting it. Score along all the other lines the same way. Fold the cardboard lengthwise in accordion pleats, folding the first fold forward.

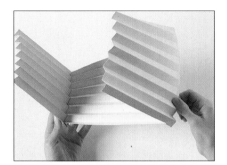

2 Open out the pleats and fold the ends upright at right angles along the broken lines. Refold the pleats. Use your fingertips to carefully pinch the outside of the folds and push in the zigzag folds to hold the corners firmly in shape.

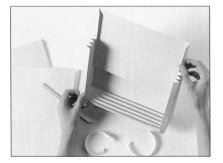

3 Refer to the diagram on p.139 to cut seven dividers from thin cardboard. Score along the broken lines with the back of a scissor blade. Apply double-sided tape along the tabs. Bend the tabs backward. Slip each divider into each division. Peel off the tape backing paper and press the tabs onto the pleats.

4 To make the cover, cut two pieces of mat board for the front and back 7¾ x 10¼ inches and a strip for the spine 1½ x 10¼ inches. Now cut colored paper or giftwrap 11¾ x 19 inches. Place the paper right-side down and position the front and back on top ¾ inch inside the outer edges. Position the spine with a ¼-inch gap between the front and back.

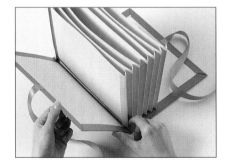

5 Stick the boards to the paper with spray glue. Stick the paper edges over the boards with double-sided tape. Glue the ends of two 10-inch lengths of ribbon centrally to the ends of the cover. Apply double-sided tape to the outer edges of the front and back dividers and pleats. Squeeze the lower pleats together. Peel off the backing tapes and stick to the front. Repeat for the back cover.

Stationery trays

KEEP STATIONERY neat and orderly by filing it in these shallow plastic trays. These desk accessories are designed with a V-shaped cut in the front that will give easy access to paper, envelopes, and a small notepad. Many hobby and craft stores stock this thin plastic, which is very strong and lightweight, yet also extremely flexible.

1 Draw the stationery tray on paper following the diagrams on p.139. Tape a piece of thin opaque plastic on top. Use a craft knife to cut out the tray, but do not cut the V-shape at this stage. Mark the dots ⅝ inch inside the corners.

2 Score along the broken lines with the back of a scissor blade so that you dent the plastic instead of cutting it. Bend the tray backward along the scored lines. Open the front out flat again and cut the V-shape.

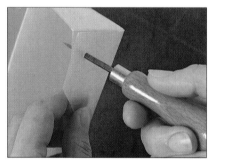

3 Refold the tray, tucking the tabs inside and matching the short edges to the folded tabs. Use an awl or scissor points to make a hole at the dots ⁵⁄₁₆ inch in diameter through the tray and tabs.

4 Cut the prongs of four large metal paper fasteners to ½ inch long. Insert the fasteners through the holes and thread on a washer, if supplied. Splay the prongs open inside the tray.

Diagrams and templates

THE FOLLOWING pages show the diagrams and templates needed to complete many of the projects. The diagrams are constructed from measurements. Use a drawing square to draw the angles accurately. Concise drawings showing how to assemble some of the wood projects have also been included as a reference guide.

All the templates are reduced in size. Using the percentage specified, enlarge each template on a photocopier, then cut it out to use as a pattern.

Hanging bag organizer (pages 42–43)

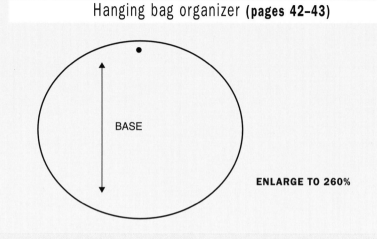

BASE

ENLARGE TO 260%

Bottle bag (pages 44–45)

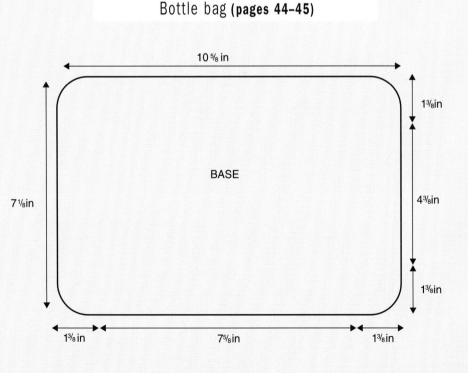

10 ⅝ in

BASE

1⅜in

4⅜in

1⅜in

7⅛in

1⅜ in

7⅝in

1⅜ in

CD stack and speaker stands (pages 50–51)

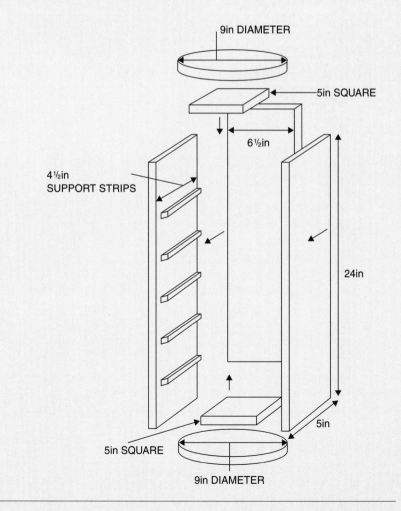

9in DIAMETER

5in SQUARE

6½in

4½in
SUPPORT STRIPS

24in

5in

5in SQUARE

9in DIAMETER

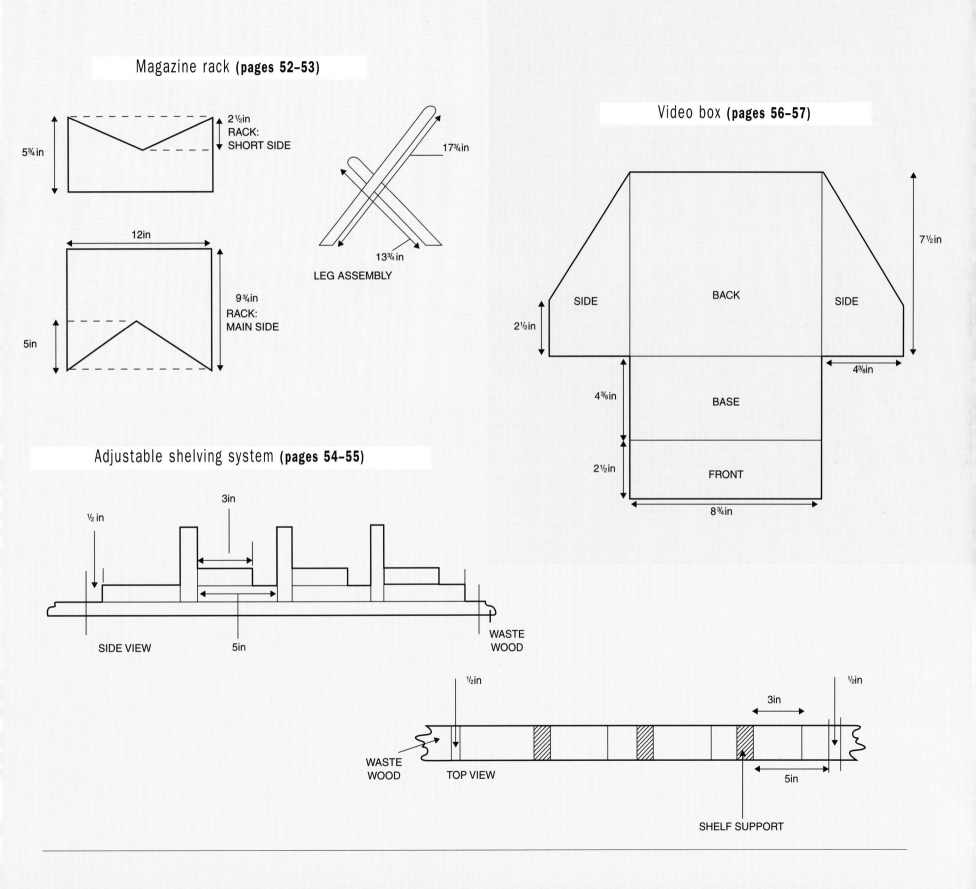

Magazine rack (pages 52-53)

2½in
RACK:
SHORT SIDE

5¾in

12in

9¾in
RACK:
MAIN SIDE

5in

17¾in

13¾in

LEG ASSEMBLY

Video box (pages 56-57)

SIDE

BACK

SIDE

7½in

2½in

4⅜in

4⅜in

BASE

2½in

FRONT

8¾in

Adjustable shelving system (pages 54-55)

3in

½ in

SIDE VIEW

5in

WASTE
WOOD

½in

3in

½in

WASTE
WOOD

TOP VIEW

5in

SHELF SUPPORT

Storage boxes (pages 60–61)

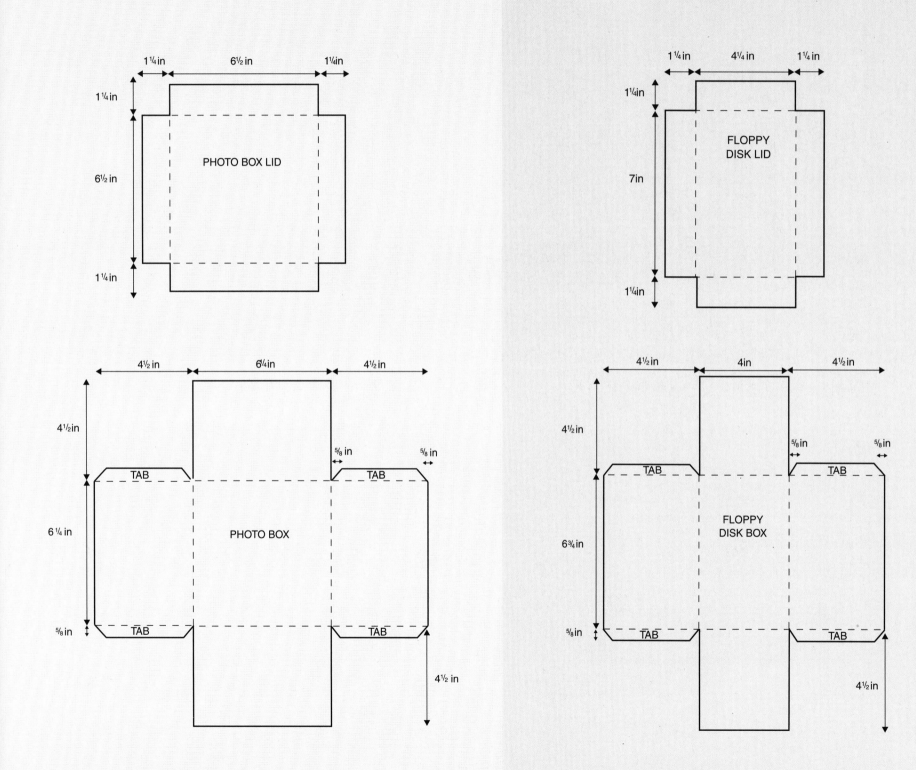

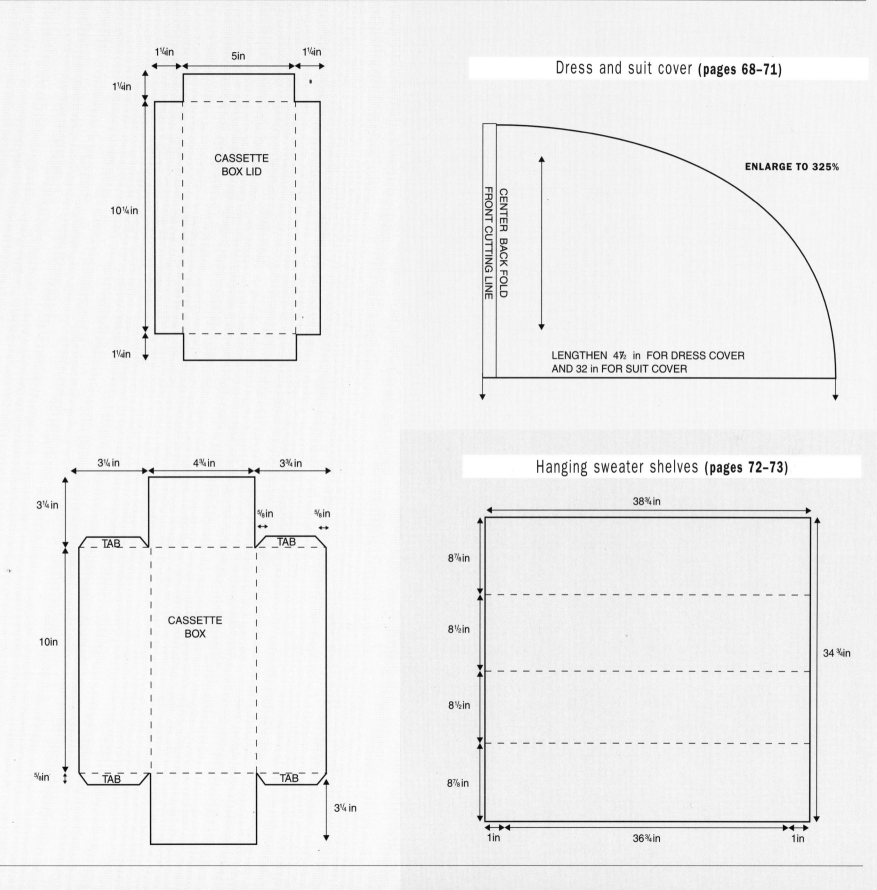

1¼in 5in 1¼in

1¼in

CASSETTE
BOX LID

10¼in

1¼in

Dress and suit cover (pages 68–71)

ENLARGE TO 325%

CENTER BACK FOLD
FRONT CUTTING LINE

LENGTHEN 47½ in FOR DRESS COVER
AND 32 in FOR SUIT COVER

3¼ in 4¾ in 3¾ in

3¼ in

⅝in ⅝in

TAB TAB

CASSETTE
BOX

10in

⅝in

TAB TAB

3¼ in

Hanging sweater shelves (pages 72–73)

38¾in

8⅞in

8½in

34¾in

8½in

8⅞in

1in 36¾in 1in

Belt and tie hanger (pages 74–75)

TOP/BASE VIEW

1½in

HOOK END

3in   2in

BASE

CENTER LINE OF ⅜ in DRILL HOLE

TOP VIEW

Blanket chest (pages 78-79)

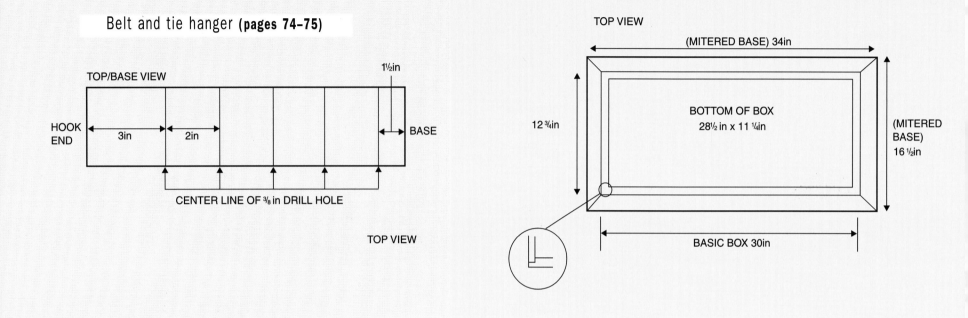

TOP VIEW

(MITERED BASE) 34in

12¾in

BOTTOM OF BOX
28½ in x 11 ¼in

(MITERED BASE)
16 ½in

BASIC BOX 30in

Clothing envelopes (pages 76–77)

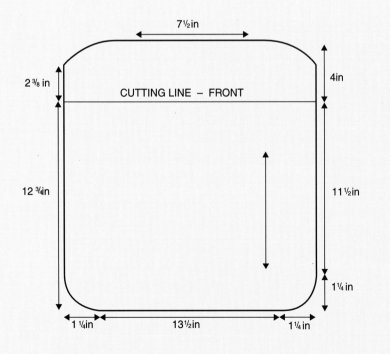

7½in

2 ⅜ in

CUTTING LINE – FRONT

4in

12 ¾in

11½in

1¼ in

1 ¼in   13½in   1 ¼in

Pegged shelf (pages 84–85)

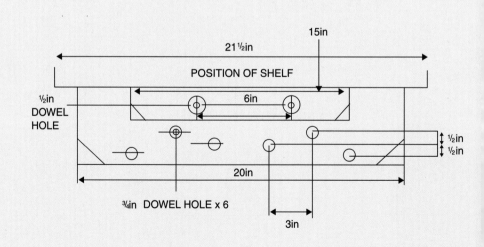

15in

21½in

POSITION OF SHELF

½in
DOWEL
HOLE

6in

½in
½in

20in

¾in DOWEL HOLE x 6

3in

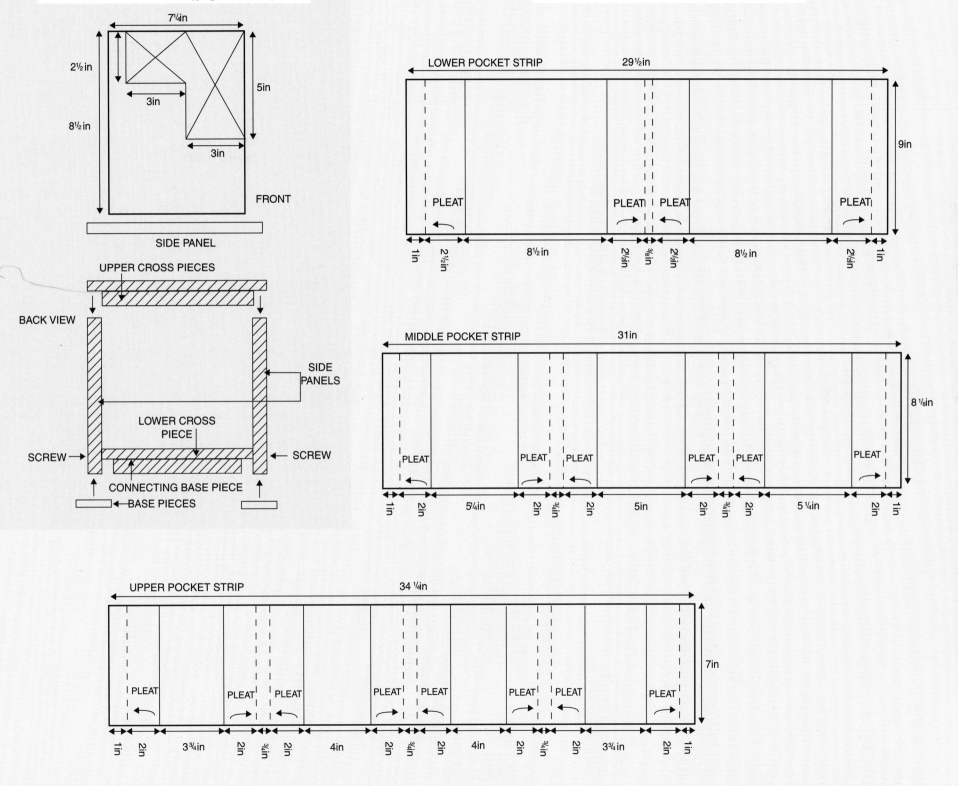

Tiered shelf unit (pages 88–89)

Toiletry pockets (pages 90–91)

Cleaning box (pages 92–93)

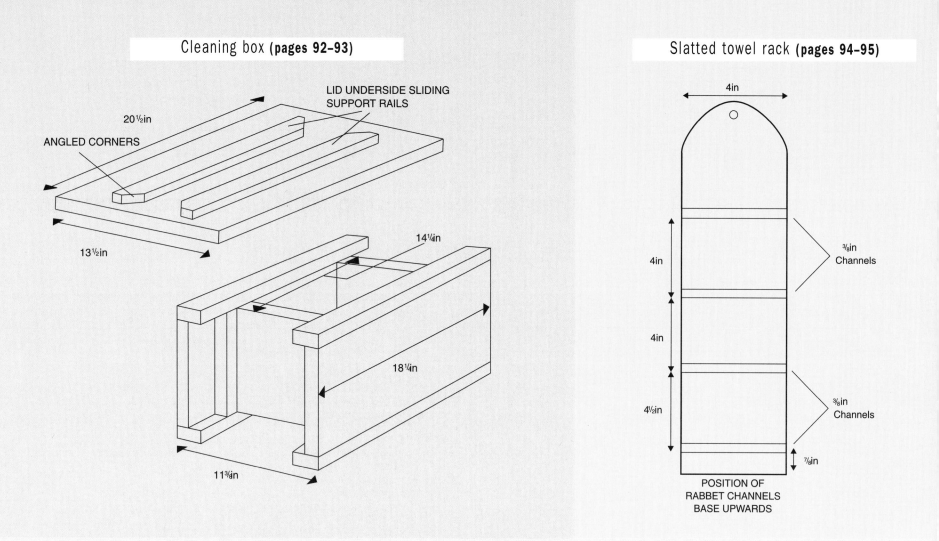

LID UNDERSIDE SLIDING
SUPPORT RAILS

20½in

ANGLED CORNERS

13½in

14¼in

18¼in

11¾in

Slatted towel rack (pages 94–95)

4in

4in

4in

4½in

⅜in
Channels

⅜in
Channels

⅞in

POSITION OF
RABBET CHANNELS
BASE UPWARDS

Fabric screen (pages 100–101)

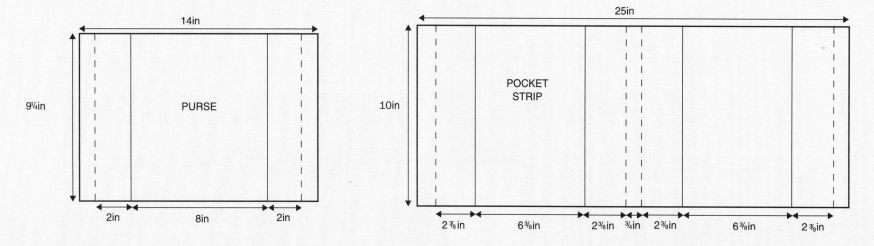

14in

9¼in

PURSE

2in

8in

2in

25in

10in

POCKET
STRIP

2⅜in

6⅜in

2⅜in

¾in

2⅜in

6⅜in

2⅜in

Toy cupboard (pages 102–103)

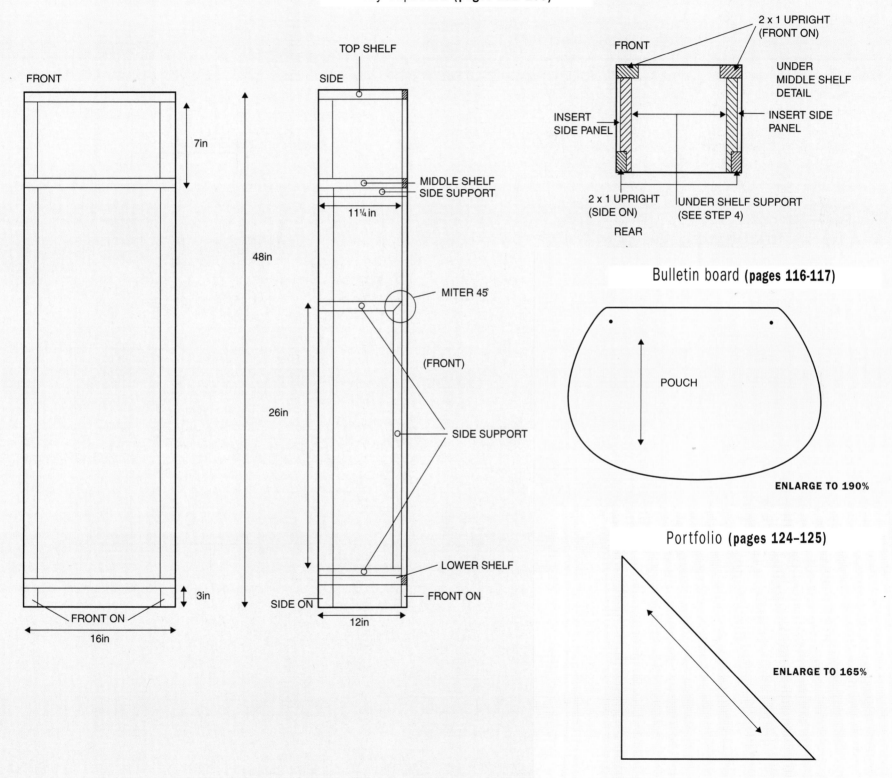

FRONT

7in

FRONT ON

16in

TOP SHELF

SIDE

MIDDLE SHELF
SIDE SUPPORT

11¼ in

48in

MITER 45°

(FRONT)

26in

SIDE SUPPORT

LOWER SHELF

SIDE ON

FRONT ON

12in

3in

FRONT

2 x 1 UPRIGHT
(FRONT ON)

UNDER
MIDDLE SHELF
DETAIL

INSERT
SIDE PANEL

INSERT SIDE
PANEL

2 x 1 UPRIGHT
(SIDE ON)

UNDER SHELF SUPPORT
(SEE STEP 4)

REAR

Bulletin board (pages 116-117)

POUCH

ENLARGE TO 190%

Portfolio (pages 124–125)

ENLARGE TO 165%

Corrugated-plastic boxes (pages 122–123)

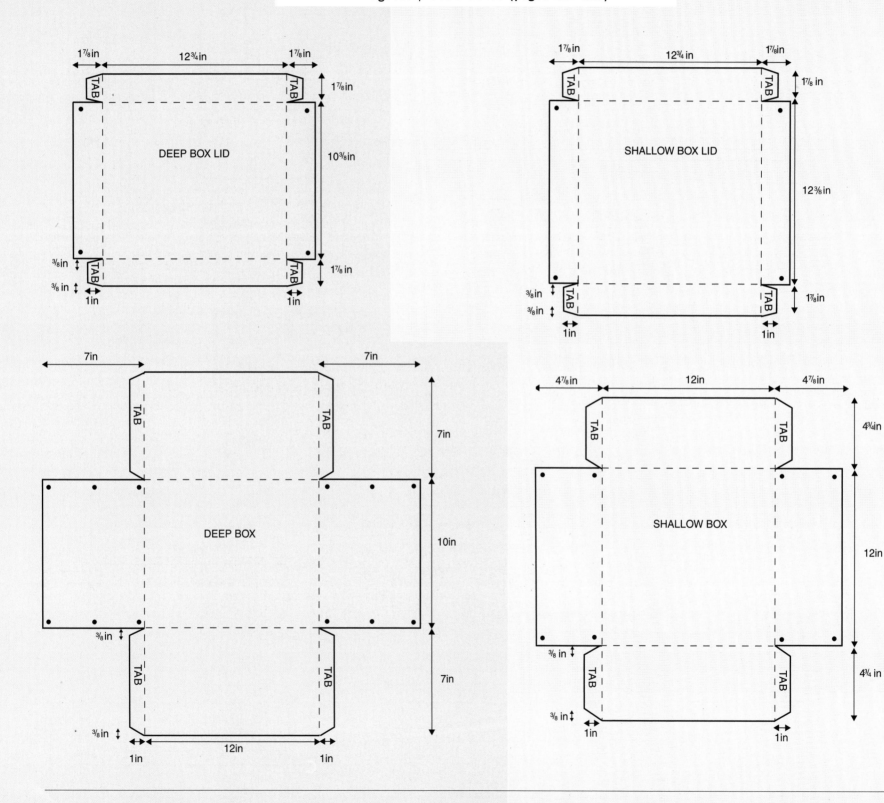

Accordion file (pages 126-127)

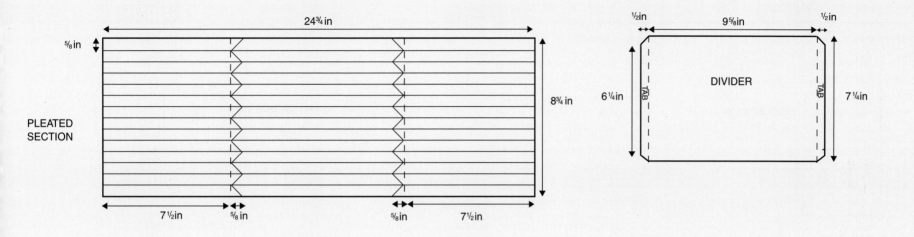

Stationery trays (pages 128–129)

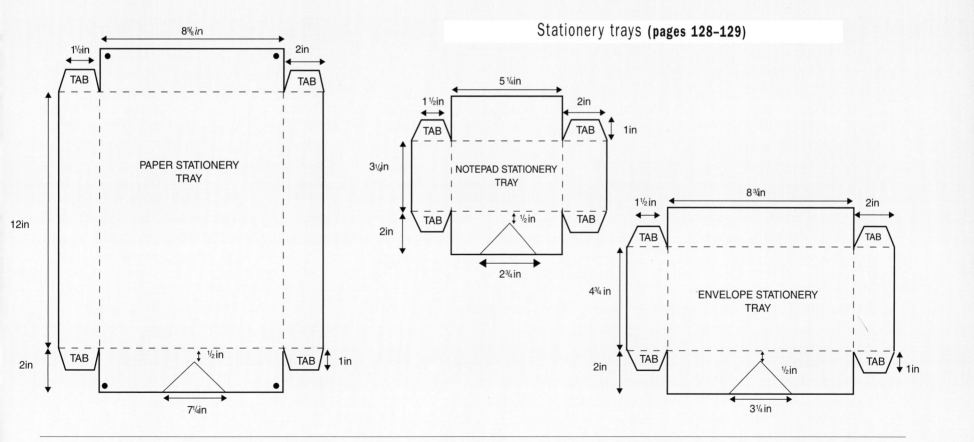

Index